Introduction to Multivariate Calibration

Alejandro C. Olivieri

Introduction to Multivariate Calibration

A Practical Approach

 Springer

Alejandro C. Olivieri
Universidad Nacional de Rosario
Instituto de Química Rosario - CONICET
Rosario, Argentina

ISBN 978-3-030-07302-2   ISBN 978-3-319-97097-4   (eBook)
https://doi.org/10.1007/978-3-319-97097-4

This Springer imprint is published by the registered company Springer Nature Switzerland AG
The registered company address is: Gewerbestrasse 11, 6330 Cham, Switzerland

To my late sister Cristina, wherever she is.

Foreword

The content of this book, written by Prof. Alejandro Olivieri, covers practical and fundamental aspects of multivariate calibration, setting the focus on first-order calibration. It is important to remark that multivariate calibration has become a crucial topic in the analytical world, nowadays being adopted in labs for solving complicated analytical problems, including those found in environmental, biochemical, agro-industry, and food analysis, among other applications.

The book has been divided in 13 chapters, starting in Chap. 1 with an introduction of what chemometrics and multivariate calibration represent in the analytical world. In this chapter, basic concepts are introduced. Then, the following chapters show the evolution that first-order calibration has experienced from the simplest and original methods to the latest algorithms, including artificial neural networks to model nonlinear systems. In addition, several chapters are devoted to explore important subjects as the optimum number of latent variables, comparison of multivariate models, data preprocessing and analytical figures of merit, and topics of supreme importance in the analytical field.

Practical aspects of multivariate calibration are discussed introducing interesting examples. Most of the experimental data used in examples and exercises correspond to methods developed by the author's research group and collaborators. In addition, the examples are intended to guide analytical chemists in their work, appealing to mathematics when it is strictly necessary and showing very smart schematic representations to understand concepts involved in these powerful multivariate chemometric tools.

To follow the explanations given in the chapters dealing with the applications, a free graphical interface software, namely, the MVC1 MATLAB routines, is presented and suggested. The software opens a very interesting scenario, as the user should be able, after the reading of this book, to exploit the usefulness of the available tools to perform calibrations with his/her own laboratory data. Particularly useful in this context is the last chapter with solutions to the homework suggested along the practical application chapters.

Finally, this book may be considered to fill a gap in the subject, with a smart treatment of the advantages and practical limitations of first-order calibration, by a combination of both fundamentals and practice, and provides free software, developed by the author, to use the most popular available approaches to deal with current analytical data.

Héctor Goicoechea
Cathedra of Analytical Chemistry
Faculty of Biochemical and Biological
Sciences, University National of Litoral
Santa Fe, Argentina

Preface

Multivariate calibration is becoming popular in analytical chemistry. Some branches of industry have used it for years; others are gradually incorporating it as the necessary tools, both theoretical and experimental, are disseminated in the academic and industrial fields. Precisely the objective of this book is to contribute to the diffusion of the discipline while providing complementary reading material for undergraduate or postgraduate courses on multivariate calibration in chemistry-related careers.

Every book on multivariate calibration faces a dilemma. Should deep mathematical concepts be employed, as those required for the development of the main tools of the discipline, or should the mathematics be kept at a minimum, describing only in a qualitative manner the different calibration techniques? The following anecdote illustrates the issue.

A journalist once interviewed a physics professor, enquiring about the relativity theory. The professor tried to explain the fundamentals of the theory, using uncommon terms such as *geodesics* and *tensors*. Thus the journalist begged for the use of more understandable terms. The professor then told a story of cowboys firing guns on a moving train and spoke about the speed of the bullets in relation to the train and to the platform.

– Good! –exclaimed the journalist– now I understand.
– Yes, but this is not the relativity theory –said the professor.

The same issue is apparent here: how much mathematics and how much qualitative text to include. Exaggerating the mathematics carries the risk of making the book difficult to understand. Reducing it to zero, on the other hand, leads to a symmetrical loss in the chemometrics. Finding the right balance in a book like this one may be a lost cause, but it is worth trying.

Regarding a chemometrics book introducing complex concepts in a simple manner, an editorial comment used the following words in Latin: *veluti pueris absinthia taetra medentes cum dare conantur, prius oras pocula circum adspirant mellis dulci flavoque liquore*. This is an advice of the poet and philosopher Lucretius to orators:

> when the topic is tough, behave as physicians seeking to give a draught of bitter wormwood to a child: first smear some honey along the edge of the cup.

The reader might be left with this same sensation with this book.

Rosario, Argentina   Alejandro C. Olivieri

Acknowledgments

To write a book like this one, it is important to have collaborated with Argentinean and foreign scientists over almost two decades and to have taught on the subject in universities and private laboratories. Scientific knowledge helps one to comply with the formalities of the discipline, and teaching practice leads one to recognize the need for printed material about an intrinsically difficult topic. It is therefore appropriate to thank the National University of Rosario and the National Scientific and Technical Research Council (CONICET) for allowing us to develop in the double role of teachers and researchers. That is not all. For this project to come true, other crucial ingredients are required: a brother-in-law (Raul), an experienced editor and excellent critic, and a wife-scientist (Graciela) with patience for consulting and discussion. For them, special thanks.

Contents

Abstract

The relationship between univariate, multivariate, and multi-way calibrations is discussed, with emphasis in the analytical advantages which can be achieved in going from simple to more complex data structures.

1.1 Chemometrics: What's in a Name?

An old joke says that statistics is a very useful discipline, because it keeps statisticians employed. We could say the same about chemometrics; chemometricians would be unemployed without it. In a similar vein, a well-known analytical chemist, Charles N. Reilley (1925–1981), said with some humor that *analytical chemistry is what analytical chemists do*. His definition of analytical chemistry intended to overcome the identity crisis faced by the discipline, which involves a number of activities coming from other well-defined scientific fields, such as chemistry, physics, mathematics, and statistics.

We could also affirm that chemometrics is what chemometricians do, because this discipline faces a similar problem, living in a border line between other traditional fields, in this case chemistry, mathematics, and statistics. Within the name of the field itself we could find, in principle, the heart of chemometrics: *chemo* refers to chemistry, and *metrics* to measurement, data processing and interpretation using statistical and mathematical models. The particle *metrics* appears in other interdisciplinary fields such as biometrics, qualimetrics, even psychometrics, where it plays the same role as in chemometrics, but complements biological sciences, quality control, and psychology.

In a broad sense, the above definition implies that fitting an analytical data set to a straight line, estimating by least-squares the slope and intercept, is a chemometric activity. Why not? I suspect that some of my colleagues will disagree with this

© Springer Nature Switzerland AG 2018
A. C. Olivieri, *Introduction to Multivariate Calibration*,
https://doi.org/10.1007/978-3-319-97097-4_1

assertion. In my opinion, it is difficult to establish how complex the data and model should be for considering the task a genuine chemometric activity. Does a line separating a *light* chemometrics from a true one exist? I doubt it.

In any case, the reader will find that the type of data and mathematical models described in this book are of a different nature in comparison with the fitting to a calibration line in classical analytical chemistry. The simple fact of measuring a set of numerical data for each experimental sample, instead of a single number (as in classical analytical calibration), opens a door to a new universe. It implies a new way of approaching the analytical problem and carries surprising analytical potentialities. The level of complexity of these activities is certainly higher than the classical fitting to the calibration line, although deep down all mathematical models of chemical data belong to chemometrics, from the simplest to the most complex ones.

Sadly, it is likely that entire branches of chemometrics constitute an unknown world to most chemists, in particular to analytical chemists. However, the specific field here described is today of utmost importance in industry (agriculture, food, chemical, textile, oil, pharmacy, biomedical, etc.), and also in basic scientific research. The main application of the methods here described lies in the possibility of replacing traditional analytical methodologies by alternative ones based on the combination of optical, electrical, and other instrumental measurements. This would avoid the use of toxic solvents, considerably decreasing energy, cost and waste, reducing the time of analysis, and performing non-invasive, remote and automatic detection. These premises are, in general terms, in agreement with the new trends to a sustainable or *green* analytical chemistry (De la Guardia and Garrigues 2012).

1.2   The Proof Is in (Eating) the Pudding

The best way to illustrate the implications of multivariate calibration is by setting an example. The analysis of glucose in the blood of diabetic patients (more than 500 million worldwide) requires the extraction of a drop of blood, and the use of auxiliary reagents on a disposable strip which is employed for the indirect electrochemical determination of the analyte. One could in principle replace this procedure by a non-invasive one, based on illuminating the skin with near infrared (NIR) light (NIR is the electromagnetic radiation in the spectral range from 2500 to 10,000 nm, which is adjacent to the red region of the visible spectrum). Such a device would register the absorption spectrum of the dermis (placed at ca. 0.01 mm under the surface when the skin is thin, as in the forearm or inside the lips) (do Amaral and Wolf 2008; Vashist 2013) (see Fig. 1.1). The NIR spectrum contains information on the absorbing chemical species which are present in blood, glucose among them. However, the relative intensity due to glucose is significantly smaller than other constituents, such as water, fat, and proteins (Malin et al. 1999). Moreover, the NIR signal is affected by physicochemical parameters such as changes in body temperature, blood pressure, and skin hydration. This should give an idea of the challenges faced by any mathematical model aimed at estimating the concentration of glucose in blood.

Incident NIR beam   Reflected NIR beam

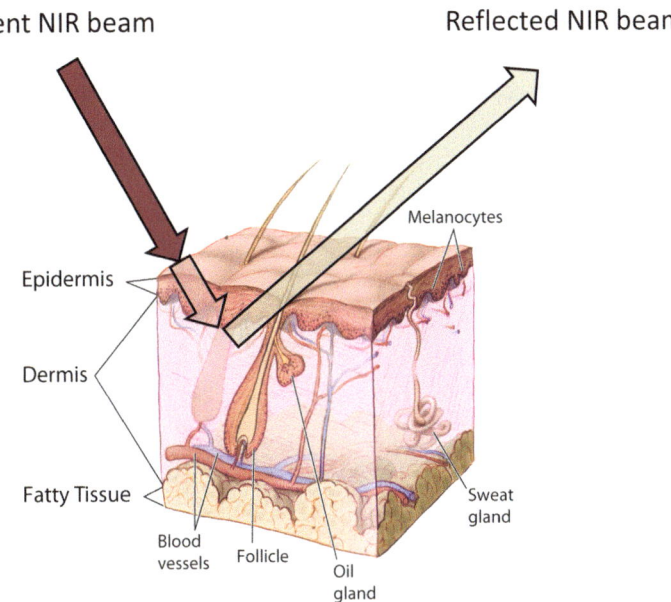

Fig. 1.1 Scheme illustrating how the glucose level can be measured in human blood by means of near infrared spectroscopy. A beam of NIR light is directed to the skin, penetrates to the dermis, and reflects back to a detector. The reflected beam contains information on the absorption spectra of all dermis constituents, including glucose. A mathematical model would then allow one to estimate the concentration of glucose. Adapted from https://commons.wikimedia.org/wiki/File:Anatomy_The_Skin_-_NCI_Visuals_Online.jpg, Bliss D, via Wikimedia Commons

If this application were possible, one could go further, developing a NIR device capable of constantly monitoring the glucose level, in real time and in a non-invasive manner, coupled to a small insulin bomb, which would introduce in the blood the amount of the hormone necessary to maintain a safe glucose concentration. This artificial pancreas would be similar to those already existing ones, which measure glucose by the traditional electrochemical method. Considering the importance of diabetes, this could be the single most relevant contribution of chemometrics for improving the human life conditions on earth.

1.3   Univariate and Multivariate Calibration

We now enter a more technical aspect: definitions. Classical analytical calibration of a single constituent is known as univariate calibration, because it is based on the measurement of a single number or datum for each experimental sample. As analytical chemists well know, this calibration requires that the measured signal be selective with respect to the analyte of interest. This sometimes implies complex operations designed to free the analyte from interfering agents which might be present in the test samples.

Fig. 1.2 Scheme illustrating how univariate calibration operates. The signals measured for the calibration samples and for the unknown sample are processed by means of a simple model (the fitting to a straight line). The result is the estimation of the analyte concentration in the unknown sample

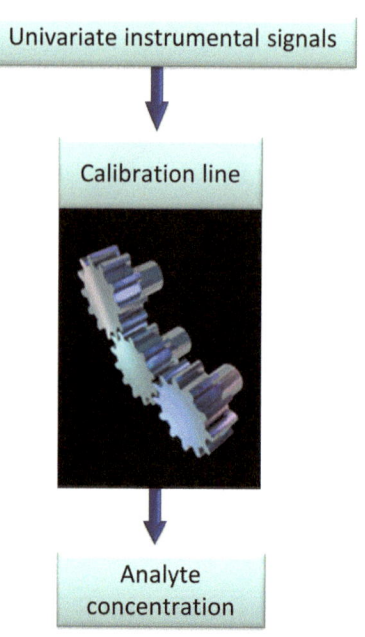

As an example, let us consider the analysis of L-malic acid in wine. Malic acid is a very important constituent; if there is not enough, the wine will taste *flat* and will be more susceptible to spoilage. If there is too much, the wine will taste *green* or *sour*. Thus, it is important for the winemaker to control the amount of malic acid present. Since L-malic is just one of the many organic acids present in wine, it is not a simple task to be able to measure its concentration by univariate calibration methods. In fact, to determine L-malic acid, it is oxidized to oxaloacetate in a reaction catalyzed by L-malate-dehydrogenase, in the presence of nicotinamide-adenine-dinucleotide (NAD). In the presence of L-glutamate, oxaloacetate is transformed in L-aspartate, in a reaction catalyzed by glutamate-oxaloacetate-transaminase. The formation of NADH (the reduced form of NAD) in this reaction is measured by the increase in absorbance at 340 nm and is proportional to the amount of L-malic in the wine sample. As can be appreciated, a considerable experimental effort is directed to isolate the analyte (L-malic) from the interferent agents, or to force it to react in a specific manner so that only the analyte generates a signal. This is the only way in which univariate calibration can be applied to the determination of specific analytes in complex samples. Once the analyte of interest is free from interferences, the univariate calibration process can be illustrated as in Fig. 1.2.

In multivariate calibration, on the other hand, several data are measured for each sample (today the number of data may be in the order of thousands or millions). In most of the examples discussed in this book, the data come from spectral absorption in the infrared or UV-visible regions, or from fluorescence emission. In these cases, the analyst measures a spectrum for each sample, i.e., a set of numbers (absorbances, reflectances, or emission intensities at multiple wavelengths) that can be arranged in

Fig. 1.3 Scheme illustrating the process of multivariate calibration. Signals for calibration and unknown samples (typically spectra or vectors of multiple signals) are processed by an appropriate chemometric model. The result is the estimation of the concentration of the analyte in the unknown sample

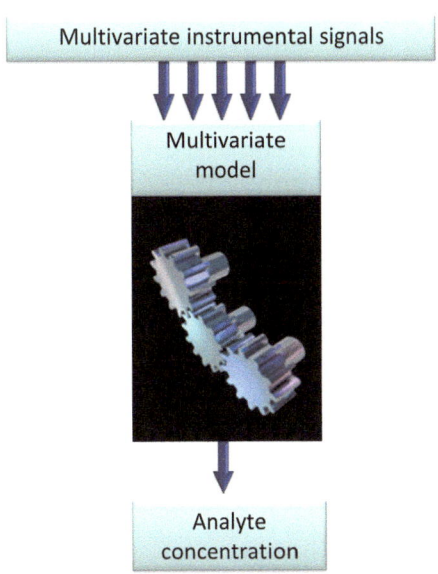

the form of a vector. For historical reasons, multivariate calibration deals mostly with spectroscopic data, but there is no reason why the analysis cannot be extended to other types of multivariate instrumental measurements. In fact, applications are known in electrochemistry (voltammetric or frequency traces, electrical sensor arrays), chromatography (depending on the type of detector, which is the device that measures the signal), etc. In the specific case of the determination of L-malic acid in wine samples, this can be reliably performed after recording a single NIR spectrum, and processing the latter with a suitable multivariate model. As a bonus, the NIR spectrum of wine contains information on other constituents, so that from a single spectrum one could in principle determine not only L-malic acid, but also other target properties as well, such as total sugars, pH, total and volatile acidity, glucose/fructose ratio, ethanol, etc. Using UV-visible spectroscopy and multivariate calibration, on the other hand, the wine content of more than 25 different phenolic compounds can be simultaneously measured in a matter of seconds (Aleixandre-Tudo et al. 2018).

In multivariate calibration, the process of determining the concentration of an analyte is similar to univariate calibration, as shown in Fig. 1.3. Notice that data processing by linear regression is replaced by a suitable multivariate model.

Today, the possibility exists of measuring data which can be grouped into more complex mathematical objects than a vector for each sample, e.g., a matrix. In the next section, we will explain some of the properties of more complex data, but our prime focus is on vectorial data.

The sea change from univariate to multivariate calibration is revolutionary. It is so in conceptual terms, but also in practical terms, judging from the variety of real-world applications that can be developed. It involves a significant change in the way of thinking the analytical experiment, which we will try to explore in the remainder of this book.

1.4   Orders and Ways

The title of this section seems to correspond to a lesson in human behavior rather than to one in chemometrics. It is not: the *order* is a property of the instrumental data for a single experimental sample, and the *way* is a property of data for a sample set. They characterize the type of mathematical objects that can be built with the data.

For example, if a single number is measured for a sample (the absorbance at a single wavelength), the order of this datum is zero. If a spectrum is measured, the absorbances can be collected into a column vector; these data are of order one or first-order. If a data table is acquired for each sample, the mathematical object is a matrix, whose order is two. And so on. The terminology has been taken from tensor algebra, where a number is a zeroth-order tensor, a vector is a first-order tensor, etc.

The number of ways characterizes the object built with the instrumental data measured for a set of samples. In the classical univariate analytical calibration, the data for a sample set generate a vector. This set is said to have only one way (the sample way). If a vector (spectrum) is measured per sample, a set of samples would result in a set of vectors, which can be accommodated side by side forming a matrix. This matrix is said to have two ways: the spectral and the sample way. When measuring data matrices for each sample, a three-dimensional object with three ways can be built. For example, matrices obtained from a liquid chromatograph with diode array detection may generate an array with three ways: the chromatographic elution time, the spectral, and the sample way.

More complex arrangements can be envisaged: three-dimensional data for a single sample lead to four-way data for a sample set, and so on. Figure 1.4 shows the progression of data from zeroth- to second-order, and from one- to three-way data.

The classical univariate calibration is equivalent to zeroth-order or one-way calibration, although it is not usually called in this manner. We here deal with first-order or two-way calibration, though the former name is probably the most

Fig. 1.4 Hierarchy of mathematical objects that can be built with the measured instrumental signals. (**A**) The order for a scalar, a vector, and a matrix for a single sample. (**B**) The number of ways for a vector, a matrix, and a three-dimensional array for a sample set

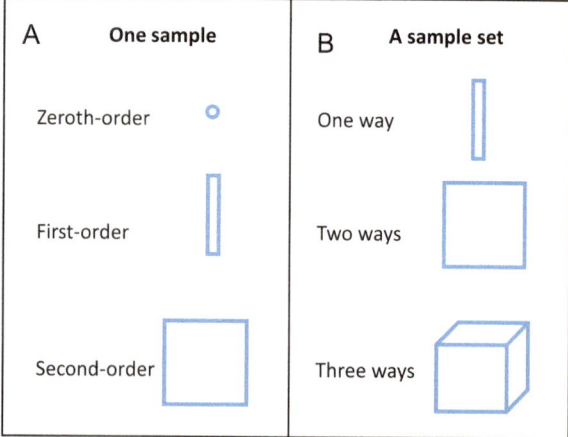

popular one. More complex mathematical objects give rise to second-order or three-way calibration, third-order or four-way calibration, etc. As a final detail, multivariate calibration includes two or more ways, i.e., from first-order and beyond. Three-way calibration and further are called multi-way calibrations (Olivieri and Escandar 2014).

We have already described the different first-order data which can be measured experimentally, giving rise to first-order calibration. Second-order data can be measured in two general manners. In a single instrument (e.g., a spectrofluorimeter), one may register fluorescence excitation-emission matrices. These are data tables in which one of the ways is the excitation wavelength and the other one is the emission wavelength (Escandar et al. 2007). Another popular form of recording second-order data is to couple two instruments in tandem: a liquid chromatograph with a diode array detector generates second-order data. In these data tables, one way is the elution time and the other one the absorption wavelength (Escandar et al. 2007). Figures 1.5 and 1.6 show three-dimensional plots for excitation-emission fluorescence and chromatography with spectral detection, respectively. These figures are also known as second-order *landscapes*, because they resemble their natural counterparts.

Third-order data can also be measured in a single instrument, by registering excitation-emission fluorescence matrices as a function of time, for example, when following the kinetic evolution of a reaction (Escandar et al. 2007). In this four-way calibration, the ways are excitation wavelength, emission wavelength, reaction time, and sample. One could also connect three instruments in tandem: in comprehensive bidimensional chromatography with spectral detection, two chromatographic columns are coupled to a multivariate detector. There are examples involving liquid

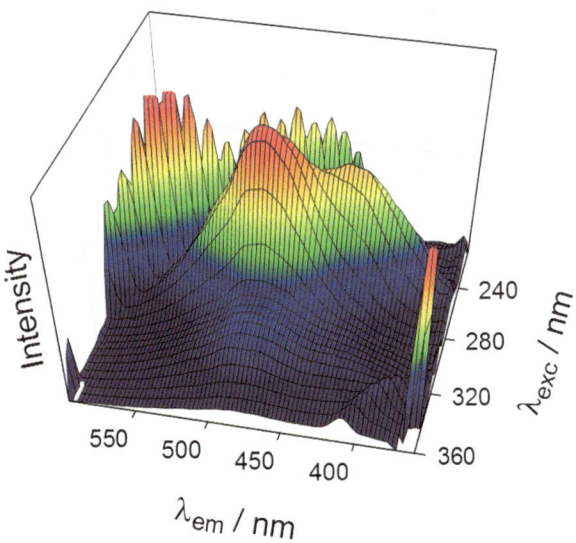

Fig. 1.5 Three-dimensional plot of an excitation-emission fluorescence matrix, showing how the emission intensity varies as a function of the excitation and emission wavelengths (labeled as λ_{exc} and λ_{em}, respectively)

Fig. 1.6 Three-dimensional plot of a matrix from chromatography with UV-visible spectral detection, showing how the absorbance varies as a function of elution time and absorption wavelength

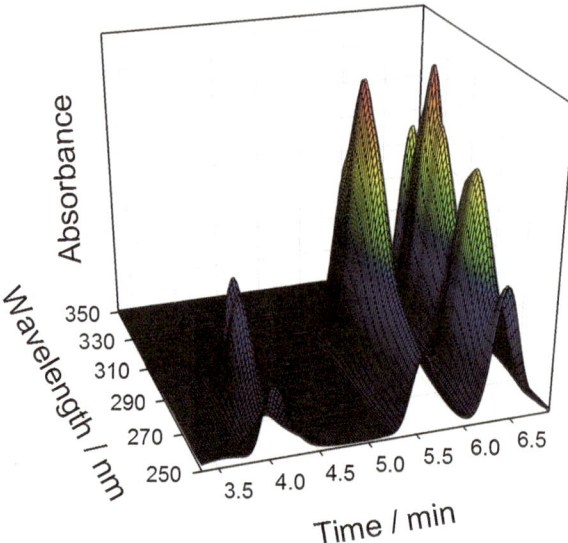

Fig. 1.7 Third-order data: temporal changes of excitation-emission fluorescence matrices (in the form of contour maps) while a chemical reaction generating fluorescent products progresses

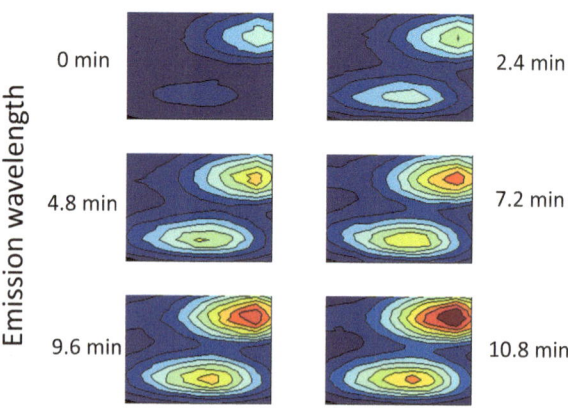

chromatography with UV-visible diode array detection (Escandar et al. 2014), and gas chromatography with mass spectrometric detection (Escandar et al. 2014).

Figure 1.7 shows how excitation-emission fluorescence matrices (represented by their contour maps) vary with the reaction time, when fluorescent products are generated while the sample constituents react with an appropriate reagent. It is not possible to plot the complete data array for a given sample, because four dimensions would be required (fluorescence intensity, excitation wavelength, emission wavelength, and reaction time).

There is no limit, in principle, for increasing the order of the data, but experimental limitations appear to exist for the development of further multi-way systems, and only a few applications have been published (Escandar et al. 2014).

1.5 Why Multivariate Calibration?

The change of attitude taking place in going from univariate to multivariate data, as regards analytical calibration, is primarily related to the role assigned to the interferents. The International Union of Pure and Applied Chemistry (IUPAC) mentions the interference issue in its definition of selectivity: *the extension that a method can be used to determine individual analytes in mixtures or matrices without interference from other components of similar behavior* (Vessman et al. 2001). In classical analytical chemistry, we are used to consider that any substance producing a signal similar to the analyte signal (for example, absorbing at the same wavelength) is an interferent. Possible solutions to avoid the undesired effect of the interferences are: (1) physically removing the interfering constituent by a clean-up procedure, (2) separating the constituents by chromatography or other separative techniques, (3) masking the interferent by reaction with a specific reagent which transforms the interferent in a non-responsive product, (4) using a specific reagent to transform the analyte into a product showing a signal different than the interferents, etc.

In multivariate calibration, on the other hand, the interferents are innocent unless proven otherwise. All interferents are, in principle, potential, and removing or transforming them is not required, as in univariate calibration. The effect of the interfering agents in the multivariate world can be appropriately compensated using mathematical models of the data for each sample.

How can this highly important analytical result be achieved? In the examples described in this book, i.e., in the framework of first-order multivariate calibration, the aim is achieved by inclusion of the potential interferents in the sample set employed for calibrating the model. This assertion opens a series of questions for the classical analytical chemist, which we will analyze below.

1.6 Frequently Asked Questions

If there were a FAQ window in multivariate analytical calibration, the most frequently asked questions would be the following ones.

1. Is it necessary to know all possible interferents that might be present in an unknown sample to apply a multivariate calibration protocol?

   In general, no; in most applications, the chemical identity of the interferents is not known.
2. How can one prepare a calibration sample set containing the interferents if they are not known?

The calibration set should contain a number of representative samples, containing varying amounts of the interferents, even when the latter ones are not known. Only the analyte content needs to be known in the calibration samples, either because a known amount of a pure analyte was used to prepare the samples, or because the analyte content was measured by a reference analytical technique.

3. Is it necessary to prepare a large number of calibration samples?

This is a very important question with no simple answer. In general, developers of multivariate calibration recommend the collection of a number of samples in the order of hundreds. The analyte content should be either known or measured in all of these samples. The calibration model will then be validated against another, independent set of samples of known analyte content. This validation phase allows one to qualify the model as satisfactory or not, depending on the average prediction error for the validation set of samples. The number of calibration samples may need to be increased if the average prediction error is large, until the multivariate model stabilizes, providing an acceptable prediction error.

4. Incidentally, what is an *acceptable prediction error*?

Roughly speaking, the average prediction error (in % relative to the mean calibration analyte content) can be characterized as excellent if it is less than 2%, good if in the range 2–5%, reasonable if in the range 5–10%, and poor if larger than 10%. However, there is an important factor to be taken into account: a newly developed method is supposed to compete with existing ones, not only in terms of relative prediction error, but also in terms of cost, speed, and simplicity. It is a balance among these parameters, and possibly other ones, what determines if a multivariate calibration is able to favorably compete or not. For example, if the only available method has an associated error of 15%, and the developed calibration model shows a relative error of 10%, then the latter one is no longer poor!

5. Does a multivariate calibration model last forever?

In general, no. With some instruments, particularly NIR spectrometers, there may be changes in the detector response or measurement conditions with time. On the other hand, there are no guarantees that the newly produced or collected samples will always have the same qualitative chemical composition, especially when the samples are of a natural origin. In any case, the multivariate models have the capacity of flagging these samples, which are different in composition with respect to the calibration set. This important property will be explored in the future, and is known as the *first-order advantage*.

6. What to do if new samples are outside the calibration range or contain chemical constituents which were not considered in the calibration phase?

This question refers to two independent issues. On one hand, if the qualitative compositions of the calibration and test samples are analogous, but the analyte occurs in the test samples at concentrations outside the calibration range, the model may still be useful. This will be possible if the relationship between signal and analyte concentration is linear, and the linearity extends beyond the calibration range.

On the other hand, if test samples contain new chemical constituents, one should recalibrate the model, including a number of new samples in the calibration set, all of known analyte content, to recover representativity and to allow the model to adapt itself to the new conditions. Industrial laboratories usually check the calibration with a certain frequency (once a month, a semester, a year, etc.) using a set of samples of known analyte content, meaning that the reference analytical methods should not be discarded, and should be kept for this periodic control of the model ability.

As an example, in a laboratory controlling the quality of sugarcane juice, a multivariate model was built to measure the Brix degrees (a measure of the content of carbohydrates in sugarcane) using NIR spectroscopy. A large number of sugarcane samples were employed for calibration, measuring the NIR spectra and the Brix degrees with a polarimeter (the reference method), achieving a reasonably stable model with good analytical parameters. However, a year of extreme cold weather in the cane producing region made the model unrepresentative. The set of control samples started to show poor analytical results, and this prompted for model re-calibration. The solution was to add, to the original calibration set, hundreds of new sugarcane juices from the cold season. The model stabilized again at a reasonable prediction error. Future cold weather conditions should not affect the calibration model.

The above considerations are valid for first-order multivariate calibration, which is the prime subject of this text. In second- and higher-order calibration, on the other hand, the view on the role of interferents is even more revolutionary, as will be discussed in Sect. 1.10. Industrial applications of higher-order data, nevertheless, are extremely limited today, and the field belongs to the area of basic scientific research. At least for now . . .

1.7   Near Infrared Spectroscopy: The Analytical Dream

What would an analytical chemist ask to Aladdin's lamp? Simple: to be able to determine the content of one or several sample constituents (or sample properties) in the following manners: instantaneous, remote, automatic and non-invasive, without subjecting the sample to clean-up or pre-treatment, and without using auxiliary reagents or organic solvents. Impossible?

Let us set an example: a fruit producing plant uses a device based on NIR spectroscopy and first-order multivariate calibration to measure, among other properties, the degree of fruit ripeness. This can be done in a completely automatic and non-invasive manner, without cutting the fruit to measure the content of fructose by liquid chromatography. It is, indeed, the dream of every analytical chemist: almost instantaneous analysis, non-invasive, automatic, and free from interferents. No organic solvents, extraction steps, or sample clean-up procedures are required.

How is it done? First, we must consider a fundamental fact: a NIR spectrometer allows one to measure the spectrum of the sample material near the surface of a solid sample, without pre-treatment or dissolution. The NIR radiation penetrates into the material up to a distance of the order of a few wavelengths, before reflecting to the detector. In this way, the spectrum of the NIR light registered by the detector contains information on the absorptive properties of the material composing the sample.[1] It is, in fact, the NIR absorption spectrum of the material, superimposed with a portion of the incident light beam which is dispersed by the sample to the detector. In this spectrum, the analyte absorption bands will be found, overlapped with those of the potential interferents.

After measuring the NIR spectra of hundreds or thousands of fruits, and determining at the same time their fructose content by a reference analytical method, one can build a first-order multivariate model. This model will correlate the NIR spectrum with the fructose level, and will allow one to know the fructose content in future fruit samples, which will only be analyzed by NIR spectroscopy.

Many similar applications of infrared (mid- and near-) and multivariate calibration are known, and almost every day new alternatives are developed. As commented above in the biomedical filed, the glucose content in blood can be measured by irradiating the skin where it is thin (the forearm or the interior of the lips) (do Amaral and Wolf 2008; Vashist 2013). Projects exist to measure usual biochemical parameters (cholesterol, albumin, uric acid, triglycerides, etc.) in a single drop of blood, simultaneously and with no auxiliary reagents (García-García et al. 2014; Perez-Guaita et al. 2012).

In the food industry, the measurement of oil, protein, starch, and moisture in seeds is already classical (Burns and Ciurczak 2008a). One could determine up to 20 different parameters in wine using as sample a single drop of wine (Gichen et al. 2005). Imagine the energy, cost, reagents, solvents, and analysis time that can be saved by replacing the classical methods for the determination of these properties with the tandem NIR-chemometrics. Do you need to analyze wine without opening the bottle? The Raman spectrum collected after passing a 1064 nm laser through the bottle green glass provides similar information (Qian et al. 2013). A nice example of the challenges faced by multivariate calibration of NIR data is provided in Fig. 1.8, which shows the spectra for a set of chopped meat samples (Borggaard and Thodberg 1992). The fact that the content of fat, protein, and moisture can be accurately determined in these samples from the highly correlated and almost featureless spectra of Fig. 1.8 can be regarded as something close to a miracle.

Portable NIR spectrometers have allowed one to develop applications which appear to be science fiction. Today, one could point to the wall of a mine a miniaturized NIR equipment of the size of a mobile phone, and get in the screen the average content of bauxite, so that aluminum extraction from the mine walls is optimized. Or point it to a dish of food and know the average content of sugars, lipid, calories, etc. These and other fascinating applications are perfectly feasible. The interested reader is directed to a classical book (Burns and Ciurczak 2008b) and to a very recent review on the subject (Pasquini 2018).

---

[1]The sample material actually involves the material that makes up the surface layer, up to a thickness of a few wavelengths. However, there are techniques that allow for non-invasive analysis of the sample bulk, as confocal Raman spectroscopy.

Fig. 1.8 NIR spectra of chopped meat samples, employed to build multivariate models for the non-invasive determination of the contents of fat, moisture, and protein. The data were recorded on a Tecator Infratec Food and Feed Analyzer working in the wavelength range 850–1050 nm, and are available at http://lib.stat.cmu.edu/datasets/tecator

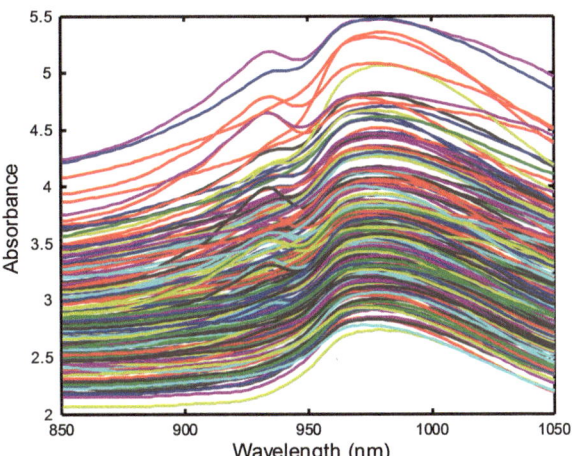

1.8   Science Fiction and Chemometrics

In the previous section, some modern applications of NIR spectroscopy and chemometrics were said to belong to science fiction. Interestingly, the famous book *I, robot* by Isaac Asimov (Asimov 1950) reflects the situation in the chapter entitled *The evitable conflict*. The following paragraphs are literally reproduced from the latter chapter.

> *The cotton industry engages experienced buyers who purchase cotton. Their procedure is to pull a tuft of cotton out of a random bale of a lot. They will look at that tuft and feel it, tease it out, listen to the crackling perhaps as they do so, touch it with their tongue, and through this procedure they will determine the class of cotton the bales represent. There are about a dozen such classes. As a result of their decisions, purchases are made at certain prices; blends are made in certain proportions. Now these buyers cannot yet be replaced by the Machine.*
>
> *Why not? Surely the data involved is not too complicated for it?*
>
> *Probably not. But what data is this you refer to? No textile chemist knows exactly what it is that the buyer tests when he feels a tuft of cotton. Presumably there's the average length of the threads, their feel, the extent and nature of their slickness, the way they hang together, and so on. Several dozen items, subconsciously weighed, out of years of experience. But the quantitative nature of these tests is not known; maybe even the very nature of some of them is not known. So we have nothing to feed the Machine. Nor can the buyers explain their own judgment.*

I, robot was published in 1950. About a decade later, multivariate calibration and near infrared spectroscopy were developed and shown to be able to do precisely what Asimov's robots appeared to be unable to do. Specifically, cotton quality can be reliably assessed by easily automatable techniques based on spectral measurements. These include the determination of cotton quality parameters such as micronaire (a measure of the air permeability of compressed cotton fibers), fineness, maturity, length, strength, uniformity, brightness, and yellowness (Zumba et al. 2017; Liu et al. 2016).

1.9   Global Properties vs. Specific Analytes

An additional characteristic of the combination spectroscopy/chemometrics is the possibility of measuring global properties of a sample, instead of quantifying individual analytes. Examples include organoleptic properties or calories in food, tenderness, freshness and shear force in meats, octane number in gasolines, distillation temperature in fuels, rheological properties of flour, textile fiber quality parameters, etc.

The determination of some global properties such as the degree of public acceptance of a food (coffee, beer, wine, chocolate, etc.) requires an adequately trained human sensorial panel. It is possible to develop multivariate calibration models of spectral data and values provided by a sensorial panel, in such a way that in the future, the acceptability of the product is only measured by spectroscopy. The success of this model greatly depends on the relationship between the concentrations of sample constituents producing spectral signal and affecting the global property to be measured. In principle, we may assume that the relationship exists, and that it is reasonable to expect that global properties such as taste or odor are directly related to the chemical constituents (and thus to their spectra).

Many published scientific papers have demonstrated the practical feasibility of calibrating multivariate models for correlating NIR spectra with organoleptic properties. An interesting example involves the study of coffee. Several coffee samples were qualified by a sensorial panel according to the following attributes: acidity, bitterness, flavor, cleanliness, body, and overall quality (Ribeiro et al. 2011). The NIR spectra of the ground coffee beans were previously recorded, and a multivariate model was built which successfully correlated the coffee attributes with the spectra. Why was the model effective? The authors showed that some bulk constituents of coffee beans are directly related to the sensorial attributes of the liquid infusion: caffeine, trigonelline, 5-caffeoylquinic acid, cellulose, coffee lipids, sucrose, and casein. This is the type of scientific work which deserves to be commended, because it not only develops a mathematical property-spectrum relation, generating a model to be applied to future samples without the need of a human panel, but also identifies the chemical constituents defining the property. Multivariate models should not be regarded as black boxes, and their physicochemical interpretability should be pursued as the ultimate source of scientific joy.

There are limitations, however, as to what can be achieved by multivariate calibration and infrared spectroscopy. The main drawback is the low sensitivity of the spectral technique. It is not possible to detect NIR signals from trace constituents in a sample; it is perfectly possible to measure bulk properties of seeds (oil, moisture, protein, starch), but not trace levels of aflatoxins. One may generate a multivariate model to measure albumin, total protein, and cholesterol in blood, all present in the concentration range 0.2–5%, but it is highly difficult to do it for creatinine or uric acid, whose levels are lower than 0.01%. In general terms, constituent concentrations of less than 1% would be difficult to be detected via NIR. If the sample properties depend on minor constituents for which the instrumental sensitivity is low, then the model will have low probabilities of succeeding.

In a student experiment published in the *Journal of Chemical Education*, a significant number of volatile constituents of banana flavor were identified by chromatography (Rasmussen 1984). They mainly include low-molecular-weight alcohols (isobutyl and isoamyl), isobutyric and isovaleric esters, and C4-C6 alkanones. A synthetic mixture was prepared with the 15 most abundant compounds in a real banana flavor, and in the same relative proportion as in the natural fruit, but … the artificial mixture did not smell as real banana! This means that the organoleptic properties may depend on chemical constituents which are present at very low concentrations, not detectable by NIR. However, other optical or electrical techniques may be more sensitive, and may improve the detectability of those minor constituents. It is a world open to scientific exploration as never before in analytical chemistry.

1.10   Multi-way Calibration and Its New Advantages

The change from univariate to first-order multivariate calibration is revolutionary in what concerns the analyst attitude with respect to interferents. Likewise, multi-way calibration (from second-order and beyond) implies a new revolutionary change, which is also related to the interferents (Olivieri and Escandar 2014).

In the multi-way universe, it is possible to calibrate an analytical system to determine an analyte in particular, without including the potential interferents in the calibration set! This may seem magical, but it is certainly not the case. This so-called *second-order advantage* is related to the mathematical properties of the objects which may be built with second- and higher-order data. Multi-way calibration may allow one, for example, to determine the analyte of interest in a complex sample containing a number of potential interferents, having calibrated the model with a minimal set of samples, only containing the pure analyte of interest (Olivieri and Escandar 2014).

The changes from zeroth- to first-order and from first- to second-order are revolutionary, but it appears that from second- to third-order and beyond the changes are not revolutionary, but only evolutionary. The general consensus is that there are no third- or higher-order advantages, so that second-order calibration and beyond all have the same second-order advantage. However, we should admit that in science the ultimate truth is never revealed, and new surprises may await analytical chemists in the future.

It is not our purpose to discuss multi-way calibration, which today is mainly confined to research laboratories, although the probability of being adopted by industry is high, given its favorable properties for the analytical work. It could mean the end of the need of cleaning samples before a chromatographic analysis, the universal solution to the baseline problem, the drastic decrease in the number of samples for building a multi-way model, etc.

One could summarize this section with a final advice to the modern analytical chemist: do it your way, but do it multi-way!

1.11 About This Book

This is the fourth of a series of books written in Spanish, which started in 2001 (Olivieri 2001) with an introduction to the field of computer programming using simple algorithms, implemented in the MATLAB environment.[2] A second one, published in 2007, described in a simple way the theory underlying multivariate calibration, trying to bring to the analytical chemist its foundations, without excessive use of linear algebra, and emphasizing the practical aspects of the discipline (Goicoechea and Olivieri 2007). Finally, a third text updated the 2007 version in a digital format (Olivieri 2017).

In this new version, the most recent developments in the field have been included. Specifically, new concepts and mathematical expressions for analytical figures of merit are introduced, a chapter is devoted to non-linear calibration using artificial neural networks, spectral pre-processing strategies are presented and their consequences are discussed, and the use of freely available stand-alone multivariate calibration software is described, which does not require the MATLAB environment to be run. The main objective of this text is to present seemingly complex material in a simple and practical manner, emphasizing the application of multivariate calibration techniques to real-world problems.

References

Aleixandre-Tudo, J.L., Nieuwoudt, H., Olivieri, A., Aleixandre, J.L., du Toit, W.: Phenolic profiling of grapes, fermenting samples and wines using UV-visible spectroscopy with chemometrics. Food Control. **85**, 11–22 (2018)
Asimov, A.: I, Robot. Gnome Press, New York (1950)
Borggaard, C., Thodberg, H.H.: Optimal minimal neural interpretation of spectra. Anal. Chem. **64**, 545–551 (1992)
Burns, D.A., Ciurczak, E.W.: Handbook of Near-Infrared Analysis, Practical Spectroscopy Series, vol. 35, 3rd edn. CRC Press, Boca Raton, FL (2008a)
Burns, D.A., Ciurczak, E.W.: Handbook of Near-Infrared Analysis, Practical Spectroscopy Series, vol. 40, 3rd edn. CRC Press, Boca Raton (2008b)
De la Guardia, M., Garrigues, S. (eds.): Handbook of Green Analytical Chemistry. Wiley, Chichester (2012)
do Amaral, C.E.F., Wolf, B.: Current development in non-invasive glucose monitoring. Med. Eng. Phys. **30**, 541–549 (2008)

---

[2]MATLAB, The Mathworks Inc, Natick, Massachusetts, USA.

Escandar, G.M., Faber, N.M., Goicoechea, H.C., Muñoz de la Peña, A., Olivieri, A.C., Poppi, R.J.: Second and third-order multivariate calibration: data, algorithms and applications. Trends Anal. Chem. **26**, 752–765 (2007)

Escandar, G.M., Goicoechea, H.C., Muñoz de la Peña, A., Olivieri, A.C.: Second- and higher-order data generation and calibration: a tutorial. Anal. Chim. Acta. **806**, 8–26 (2014)

García-García, J.L., Pérez-Guaita, D., Ventura-Gayete, J., Garrigues, J., de la Guardia, M.: Determination of biochemical parameters in human serum by near-infrared spectroscopy. Anal. Methods. **6**, 3982–3989 (2014)

Gichen, M., Dambers, R.G., Cozzolino, D.: A review of some applications in the Australian wine industry. Aust. J. Grape Wine Res. **11**, 296–305 (2005)

Goicoechea, H.C., Olivieri, A.C.: La calibración en Química Analítica. Universidad Nacional del Litoral, Santa Fe (2007)

Liu, Y., Delhom, C., Todd Campbell, B., Martin, V.: Application of near infrared spectroscopy in cotton fiber micronaire measurement. Inform. Process. Agric. **3**, 30–35 (2016)

Malin, S.F., Ruchti, T.L., Blank, T.B., Thennadil, S.N., Monfre, S.L.: Noninvasive prediction of glucose by near-infrared diffuse reflectance spectroscopy. Clin. Chem. **45**, 1651–1658 (1999)

Olivieri, A.C.: Calibración multivariada. Introducción a la programación en MATLAB. Ediciones Científicas Argentinas, Buenos Aires (2001)

Olivieri, A.C.: Calibración multivariada: Una aproximación práctica. Ediciones Científicas Argentinas, Buenos Aires (2017)

Olivieri, A.C., Escandar, G.M.: Practical Three-Way Calibration. Elsevier, Waltham (2014)

Pasquini, C.: Near infrared spectroscopy: a mature analytical technique with new perspectives – a review. Anal. Chim. Acta. **1026**, 8–36 (2018). https://doi.org/10.1016/j.aca.2018.04.004

Perez-Guaita, D., Ventura-Gayete, J., Pérez-Rambla, C., Sancho-Andreu, M., Garrigues, S., de la Guardia, M.: Protein determination in serum and whole blood by attenuated total reflectance infrared spectroscopy. Anal. Bioanal. Chem. **404**, 649–656 (2012)

Qian, J., Wu, H., Lieber, C., Bergles, E.: Non destructive red wine measurement with dispersive 1064 nm Raman Spectroscopy. Technical Note from BaySpec. www.bayspec.com (2013)

Rasmussen, P.W.: Qualitative analysis by gas chromatography. GC versus the nose in formulating artificial fruit flavors. J. Chem. Educ. **61**, 62–67 (1984)

Ribeiro, J.S., Ferreira, M.M., Salva, T.J.: Chemometric models for the quantitative descriptive sensory analysis of Arabica coffee beverages using near infrared spectroscopy. Talanta. **83**, 1352–1358 (2011)

Vashist, S.K.: Continuous glucose monitoring systems: a review. Diagnostics. **3**, 385–412 (2013)

Vessman, J., Stefan, R.I., van Staden, J.F., Danzer, K., Lindner, W., Burns, D.T., Fajgelj, A., Müller, H.: Selectivity in analytical chemistry. Pure Appl. Chem. **73**, 1381–1386 (2001)

Zumba, J., Rodgers, J., Indest, M.: Fiber micronaire, fineness, and maturity predictions using NIR spectroscopy instruments on seed cotton and cotton fiber, in and outside the laboratory. J. Cotton Sci. **21**, 247–258 (2017)

The Classical Least-Squares Model

2

Abstract

The simplest first-order multivariate model, based on classical least-squares, is discussed. Important concepts are introduced, which are common to other advanced models, such as the regression coefficients and the first-order advantage. The main limitations of the classical model are detailed.

2.1 Direct and Inverse Models

First-order multivariate models can be classified as either *direct* or *inverse*. The nomenclature refers to the manner in which the relationship between signal (in fact, multivariate signals measured at multiple sensors, or at multiple wavelengths in spectroscopy) and concentration of chemical constituents is established. In the direct models, the signal is considered to be directly proportional to the concentration, as dictated, for example, by Lambert–Beer's law in classical UV-visible spectroscopy. In the inverse models, on the other hand, the concentration is considered to be directly proportional to the signal.

How important can the difference be? In classical univariate calibration, no substantial difference appears to exist in calibrating a regression line in a direct or in an inverse fashion. However, the inverse univariate model seems to be more efficient regarding analyte prediction when the data sets are small and the noise level is high (Tellinghuisen 2000).

In multivariate calibration, on the other hand, the difference between direct and inverse models is crucial. Direct models, as we shall see in this chapter, show some advantages, but present unsolvable problems regarding the variety of analytical systems to be tackled, some of which were qualitatively discussed in the previous chapter.

© Springer Nature Switzerland AG 2018

A. C. Olivieri, *Introduction to Multivariate Calibration*,

https://doi.org/10.1007/978-3-319-97097-4_2

2.2   Calibration Phase

The simplest direct multivariate model is known as classical least-squares (CLS) (Thomas and Haaland 1988). This model has been almost abandoned today, due to its disadvantages in comparison with the inverse models. However, it shows a high pedagogical value. It is simple and intuitive, and it follows the classical Lambert–Beer's law of absorption spectroscopy, allowing one to introduce some abstract concepts that will be greatly useful when studying the inverse models.

As all analytical methods, multivariate calibration consists of a first phase, in which the calibration model is built from a set of experimental samples where the analyte contents and spectra (or other multivariate signals) are known. Some authors call this stage the *training* phase, a term related to the idea that multivariate calibration is some type of artificial intelligence, in which the model *learns* the spectra–concentration relationship, before facing the world of new and unknown samples.

For building a CLS model, it is necessary to prepare mixtures of standards of all pure chemical constituents to be used as calibration samples. The number of mixtures should be at least equal to the number of constituents, but in general analysts prefer to prepare a generously larger number of samples than constituents, because in this way the results are more precise. This is similar to univariate calibration, where several standards are employed to determine a single analyte. Notice that we distinguish between *constituents* and *analytes*. The former ones are generic chemical compounds present in typical experimental samples, although not all of them may be analytes of interest. It may perfectly happen, for example, that samples of pharmaceuticals contain five constituents, two of which are active principles (the analytes) and the remaining three are excipients (constituents but not analytes). We prefer not to use the term *component*, because the latter may be confused with abstract entities calculated by certain mathematical techniques, and do not represent, in general, true chemical constituents.

The CLS calibration phase faces two problems: (1) how many samples should be prepared for calibration and (2) which concentrations should be assigned to their constituents before preparing the calibration samples. The overall issue is part of a chemometric field called *experimental design*. The theory behind mixture design is beyond the scope of this book; we may simply detail, using common sense, that the calibration mixtures should be representative, as much as possible, of the combination of concentrations to be found in future, unknown samples. The specific values of the constituent concentrations should fulfill certain requirements, which will be apparent when the CLS model is developed in detail.

Let us assume that several (the specific number is I) standard solutions of constituents have been prepared, and the absorbances of these solutions have been measured at J different wavelengths. The corresponding instrumental responses x_{ji} (the absorbance at wavelength j of standard solution i) can be collected in a calibration matrix $\mathbf{X}$ (of size $J \times I$):

$$\mathbf{X} = \begin{bmatrix} x_{11} & x_{12} & \cdots & x_{1I} \\ x_{21} & x_{22} & \cdots & x_{2I} \\ \cdots & \cdots & \cdots & \cdots \\ x_{J1} & x_{J2} & \cdots & x_{JI} \end{bmatrix} \tag{2.1}$$

Figure 2.1 shows the matrix $\mathbf{X}$ in detail, and its relationship with the experimental spectra of the calibration samples. On the other hand, the concentrations of the N chemical constituents in the calibration set must be known. They are grouped in the calibration concentration matrix $\mathbf{Y}$ (of size $I \times N$), whose generic element y_{in} is the concentration in the mixture i of constituent n:

$$\mathbf{Y} = \begin{bmatrix} y_{11} & y_{12} & \cdots & y_{1N} \\ y_{21} & y_{22} & \cdots & y_{2N} \\ \cdots & \cdots & \cdots & \cdots \\ y_{I1} & y_{I2} & \cdots & y_{IN} \end{bmatrix} \tag{2.2}$$

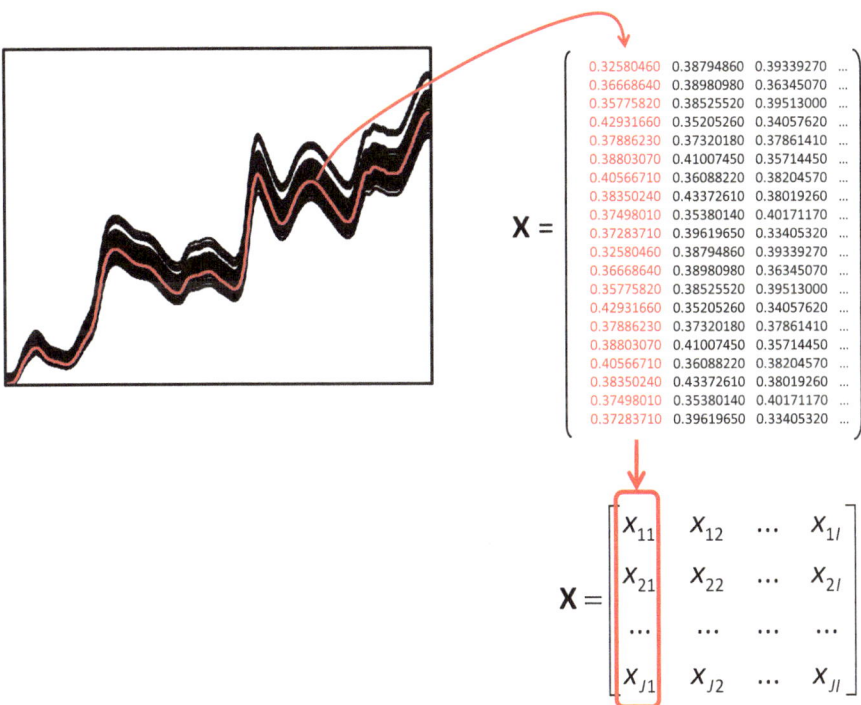

Fig. 2.1 Left, spectra registered at different wavelengths, or multivariate signals registered at different sensors. Top right, the matrix $\mathbf{X}$ in detail, showing that each spectrum corresponds to a column. Bottom right, generic representation of the matrix $\mathbf{X}$, highlighting the specific column with a red box

Figure 2.2 schematically illustrates how the constituent concentrations in the calibration set are organized in a data table or matrix.

The calibration phase is completed by assuming that Lambert–Beer's law is obeyed between absorbance and concentration, or a similar linear law relating a generic signal and concentration. The mathematical expression for the direct CLS model is the equation relating $\mathbf{X}$ and $\mathbf{Y}$ through a matrix of proportionality constants called $\mathbf{S}$ (of size $J \times N$, whose generic element s_{jn} is the sensitivity at wavelength j of constituent n):

$$\mathbf{X} = \mathbf{S}\,\mathbf{Y}^{\mathrm{T}} + \mathbf{E} \tag{2.3}$$

Figure 2.3 shows a simple schematic representation of the former equation using blocks for representing the matrices. Notice that in UV-visible absorption spectroscopy, the element s_{jn} is the molar absorptivity at wavelength j of constituent n. In general, the name *sensitivity* is preferred for this element, because it may be applied to any instrumental technique beyond UV-visible spectroscopy.

Sample	Constituent 1	Constituent 2	Constituent ...
1	10.5	0.02	...
2	7.8	0.15	...
3	4.9	0.58	...
...	...	...	...

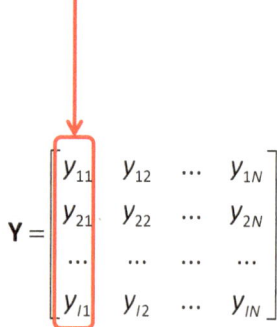

$$\mathbf{Y} = \begin{bmatrix} y_{11} & y_{12} & \cdots & y_{1N} \\ y_{21} & y_{22} & \cdots & y_{2N} \\ \cdots & \cdots & \cdots & \cdots \\ y_{I1} & y_{I2} & \cdots & y_{IN} \end{bmatrix}$$

Fig. 2.2 Top, data table with the constituent concentrations of each calibration sample. Bottom, matrix $\mathbf{Y}$ with the calibration concentrations. The red arrow and box connects the column of the table with the column of the matrix $\mathbf{Y}$

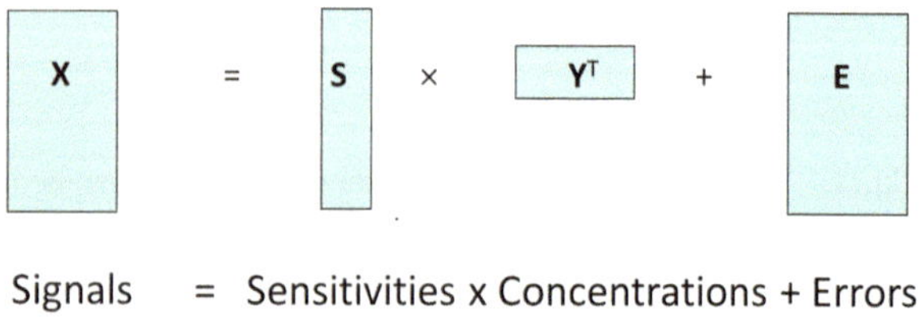

Fig. 2.3 Block scheme representing the CLS model

$$\mathbf{X} \quad = \quad \mathbf{S} \quad \times \quad \mathbf{Y}^{\mathrm{T}} \quad + \quad \mathbf{E}$$

Signals   =   Sensitivities x Concentrations + Errors

In comparison with the simple univariate law ($x = s\,y$), two differences are found in Eq. (2.3): (1) the matrix $\mathbf{Y}$ is transposed (columns are converted into rows and vice versa), as symbolized with the superscript "T," and (2) an error matrix $\mathbf{E}$ is summed to the right-hand side. The first difference is due to consistency between matrix sizes and is only formal. The second one is more important, and deserves a detailed study, as the one presented in the next section.

The CLS calibration phase consists, therefore, in the preparation of the standard solutions (mixtures of the pure constituents at known concentrations contained in matrix $\mathbf{Y}$), measurement of their spectra (contained in matrix $\mathbf{X}$), mathematical relation through Eq. (2.3) expressing the linear signal-concentration law, and finally estimation of the matrix $\mathbf{S}$. The latter matrix is the necessary link between signals and concentrations, to be applied to future samples. The next sections will deal with the details.

2.3   Model Applicability

Several requirements should be fulfilled for the successful application of the CLS model to a real analytical system. In the first place, since the total signal in the calibration samples ($\mathbf{X}$) will be modeled as a function of the concentrations of the chemical constituents producing instrumental response ($\mathbf{Y}$), it is necessary to know the concentrations of all these constituents in all calibration samples. The constituents should be the same that will occur in the future test samples. This is the main problem associated to the CLS model. Think about the determination of glucose in blood: a CLS calibration would require one to know all the chemical identities and concentrations of all the human blood constituents producing a NIR signal. This is clearly not possible.

Additionally, it is also important to recall that the CLS model relates signals to analyte concentrations, but not signals to global properties (octane number, sensorial attributes, etc.). This means that the CLS model is not applicable to the prediction of properties of samples such as foodstuff, drinks, perfumes, and fuels, a fact that further limits its application field.

When can one apply the CLS model with success? One possible application area is the analysis of pharmaceutical products, in which every constituent of a pharmaceutical form is most likely to be known, and all necessary standards are available to prepare the calibration mixtures, including active principles and excipients, and even degradation or synthetic by-products. The developed model would allow the simultaneous determination of the active principles by classical spectroscopic techniques, in the presence of excipients and other constituents, without the need of physically separating them or using chromatographic methods. Nevertheless, even in these cases inverse multivariate models might be the preferred (Goicoechea and Olivieri 1998).

2.4   Why Least-Squares? Mathematical Requirements

Calibrating a CLS model requires to estimate the matrix $\mathbf{S}$ from Eq. (2.3). We should recall that the latter equation is, in fact, the formal expression of a set of simultaneous equations with multiple unknowns. One of these specific equations, corresponding to a given element of $\mathbf{X}$, is shown in Fig. 2.4.

If the samples contain N constituents, $J \times N$ parameters should be estimated (the values of all s_{jn} elements of matrix $\mathbf{S}$) from a set of $J \times I$ equations (the number of elements of $\mathbf{X}$). In general $I > N$, and thus the problem is over-determined, i.e., there are more equations than unknowns. In this case, the usually employed criterion is to estimate $\mathbf{S}$ as the least-squares solution, that is, by minimizing the sum of squared elements of the matrix of errors $\mathbf{E}$ in Eq. (2.3). It can be shown that the least-squares solution for $\mathbf{S}$ is equivalent to finding $\mathbf{S}$ from Eq. (2.3), removing the $\mathbf{E}$ term. However, this estimation cannot be simply done by post-multiplying Eq. (2.3) by $(\mathbf{Y}^T)^{-1}$, because $\mathbf{Y}$ is not, in general, a square matrix, and non-square matrices cannot be inverted. Thus, a previous phase is required: both sides of Eq. (2.3) are first post-multiplied by matrix $\mathbf{Y}$:

$$\mathbf{X}\,\mathbf{Y} = \mathbf{S}\,\mathbf{Y}^T\mathbf{Y} \tag{2.4}$$

Notice that $\mathbf{E} = \mathbf{0}$ has been set before this operation. The product $(\mathbf{Y}^T\,\mathbf{Y})$ is a square matrix (of size $N \times N$), and post-multiplying both sides of Eq. (2.4) by the inverse $(\mathbf{Y}^T\,\mathbf{Y})^{-1}$ leads to the estimation of $\mathbf{S}$:

$$\mathbf{S} = \mathbf{X}\mathbf{Y}(\mathbf{Y}^T\mathbf{Y})^{-1} \tag{2.5}$$

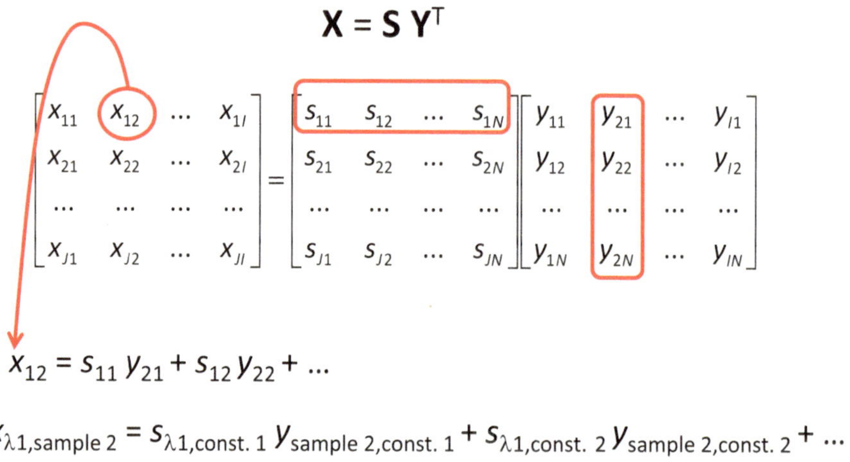

$$\mathbf{X} = \mathbf{S}\,\mathbf{Y}^T$$

$$x_{12} = s_{11}\,y_{21} + s_{12}\,y_{22} + \dots$$

$$x_{\lambda 1,\text{sample 2}} = s_{\lambda 1,\text{const. 1}}\,y_{\text{sample 2,const. 1}} + s_{\lambda 1,\text{const. 2}}\,y_{\text{sample 2,const. 2}} + \dots$$

Fig. 2.4 A specific element of the matrix $\mathbf{X}$ (x_{12}) is shown to correspond to one of the multiple equations, as the product of the first row of $\mathbf{S}$ and the second column of $\mathbf{Y}^T$ (both highlighted by red boxes). The latter product reflects Lambert–Beer's law and the concept of signal additivity, since x_{12} is given by $s_{11}\,y_{21} + s_{12}\,y_{22} + \dots$, i.e., as a sum of products of the form (absorptivity $\times$ concentration)

Equation (2.5) deserves some comments. First, it is necessary to remark that to be able to solve it, the square matrix $(\mathbf{Y}^T\mathbf{Y})$ needs to be inverted. Inversion of a matrix requires its lines (rows or columns) to be linearly independent, meaning that they should not be linear combinations of other lines. In the present case this implies, from a chemical viewpoint, that the constituent concentrations in the calibration mixtures are not correlated (for example, they do not simultaneously increase or decrease from one mixture to another). Designing a mixture set with minimal correlations is also part of the theory of experimental design. The presence of correlations in $\mathbf{Y}$ would make the determinant of the matrix $(\mathbf{Y}^T\mathbf{Y})$ null, precluding the matrix inversion.

The second comment is only formal. If we call the matrix operation $[\mathbf{Y}\,(\mathbf{Y}^T\mathbf{Y})^{-1}]$ as $\mathbf{Y}^+$, the latter is a kind of inverse of $\mathbf{Y}$ (transposed, to be precise). The literature calls $\mathbf{Y}^+$ as the *generalized inverse* of $\mathbf{Y}$. With this nomenclature, Eq. (2.5) can be written in the following compact form:

$$\mathbf{S} = \mathbf{X}(\mathbf{Y}^+)^T \tag{2.6}$$

This is the equation representing the calibration phase, providing a matrix $\mathbf{S}$ for prediction in future samples. The obtainment of $\mathbf{S}$ is analogous to the estimation of the slope of the univariate calibration line, previous to the measurement of the analytical signal for unknown samples.

2.5   Prediction Phase

In the prediction phase, an unknown sample provides J values of the instrumental signal, e.g., J absorbances at the same wavelengths at which the calibration signals were measured. The sample instrumental responses are grouped in a column vector (of size $J \times 1$) $\mathbf{x}$:

$$\mathbf{x} = \begin{bmatrix} x_1 \\ x_2 \\ \dots \\ x_J \end{bmatrix} \tag{2.7}$$

Prediction proceeds by resorting to Lambert–Beer's law applied to the unknown sample, analogously to Eq. (2.3):

$$\mathbf{x} = \mathbf{S}\,\mathbf{y} + \mathbf{e} \tag{2.8}$$

where $\mathbf{y}$ is a column vector (of size $N \times 1$) containing N elements: the concentrations of the N constituents of the unknown sample, and $\mathbf{e}$ is a vector collecting the errors of the linear model. We are again in the presence of a system of simultaneous equations with multiple unknowns: the J equations correspond to the J elements of $\mathbf{x}$, and the N unknowns to the concentrations of the N constituents contained in $\mathbf{y}$. One of such equations is $x_j = s_{j1} y_1 + s_{j2} y_2 + \ldots$, where x_j is an element of the vector $\mathbf{x}$, s_{j1}, s_{j2}, $\ldots$ are elements of $\mathbf{S}$ (sensitivities at wavelength j for each sample constituent 1, 2, $\ldots$), and y_1, y_2, $\ldots$ are the elements of $\mathbf{y}$ (concentrations of each constituent in the test sample). Again, this complies both with the signal additivity concept and Lambert–Beer's law.

The problem will be over-determined if $J > N$, which normally occurs (there are considerably more wavelengths in the spectra than sample constituents), so that the criterion of least-squares is also employed to solve Eq. (2.8) to find $\mathbf{y}$ (fixing $\mathbf{e} = \mathbf{0}$). Equation (2.8) should be first pre-multiplied by $\mathbf{S}^{\mathrm{T}}$, to obtain a square matrix in the right-hand side:

$$\mathbf{S}^{\mathrm{T}}\mathbf{x} = (\mathbf{S}^{\mathrm{T}}\mathbf{S})\mathbf{y} \qquad (2.9)$$

One could then find $\mathbf{y}$ pre-multiplying by the inverse of $(\mathbf{S}^{\mathrm{T}}\ \mathbf{S})$:

$$\mathbf{y} = (\mathbf{S}^{\mathrm{T}}\mathbf{S})^{-1}\mathbf{S}^{\mathrm{T}}\mathbf{x} \qquad (2.10)$$

We could also define the generalized inverse of $\mathbf{S}$ which would allow to find $\mathbf{y}$ pre-multiplying $\mathbf{x}$:

$$\mathbf{y} = \mathbf{S}^{+}\mathbf{x} \qquad (2.11)$$

Are there other mathematical requirements in this phase due to the need of inverting the matrix $(\mathbf{S}^{\mathrm{T}}\ \mathbf{S})$ in Eq. (2.10)? We may guess the answer: the columns of $\mathbf{S}$ (the pure constituent spectra) should not be correlated. The spectral correlation is also known as collinearity. If the constituent spectra are significantly collinear, the determinant of $(\mathbf{S}^{\mathrm{T}}\ \mathbf{S})$ will be zero or close to zero, it would be impossible (or very difficult) to find the inverse $(\mathbf{S}^{\mathrm{T}}\ \mathbf{S})^{-1}$, and the analyte concentrations will be poorly defined. The result will be a considerably high prediction error (Goicoechea and Olivieri 1998). Intuitively, if two constituents have identical spectra, it will not be possible to distinguish them. The mathematical consequence of the equivalence of analyte spectra is that the matrix $(\mathbf{S}^{\mathrm{T}}\ \mathbf{S})$ is not invertible. We always remark the importance of connecting a purely mathematical result with a qualitative observation with physicochemical meaning.

As a summary of the CLS model, after calibration with mixtures of constituents providing the matrix $\mathbf{S}$ from their spectra and concentrations, the spectrum of an unknown mixture is measured, and Eq. (2.11) provides access to the concentrations of all sample constituents, without requiring their physical separation. This analysis could not be made by classical univariate calibration, unless the constituent spectra do not overlap at all. The spectral overlapping precludes classical analysis, requiring separation methods such as chromatography, or chemometric methods such as the presently discussed one. Some authors call multivariate models *virtual chromatography*, *mathematical chromatography*, or simply with the play on words *chroMATHography*.

2.6   The Vector of Regression Coefficients

Let us suppose that a CLS model has been built for samples with various constituents, but only one of them is the analyte of interest, while the concentrations of the remaining constituents do not show any analytical interest. It is possible to isolate a portion of Eq. (2.11) which only allows one to estimate the concentration of the nth analyte in the unknown sample. This is equivalent to isolate, from matrix $\mathbf{S}^+$, the row corresponding to this analyte, and multiply it by the vector of signals for the unknown sample $\mathbf{x}$:

$$y_n = (n\text{th row of } \mathbf{S}^+)\mathbf{x} \tag{2.12}$$

A block representation of the latter equation is shown in Fig. 2.5, by extracting a specific row of the matrix $\mathbf{S}^+$, which is linked to the analyte of interest. The nth row of $\mathbf{S}^+$ once transposed (converted into a column vector) is known as the vector of regression coefficients for the nth analyte, and is represented as $\mathbf{b}_n$:

$$\mathbf{b}_n = (n\text{th row of } \mathbf{S}^+)^\mathrm{T} \tag{2.13}$$

With this latter definition, Eq. (2.12) becomes:

$$y_n = \mathbf{b}_n^\mathrm{T}\mathbf{x} = b_{1n}x_1 + b_{2n}x_2 + \ldots + b_{Jn}x_J \tag{2.14}$$

meaning that the estimated analyte concentration is the scalar product of the vector of regression coefficients by the vector of instrumental responses.

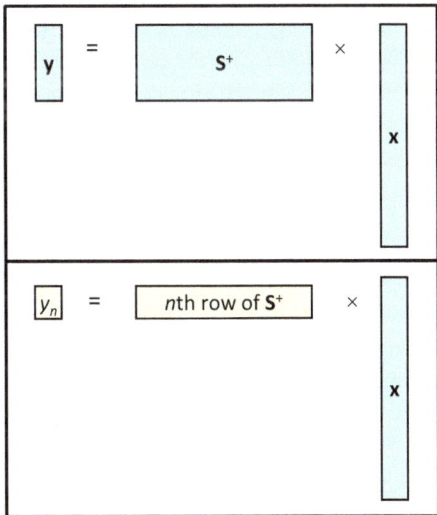

Fig. 2.5 Top, block representation of the estimation of the constituent concentrations in an unknown sample from the generalized inverse $\mathbf{S}^+$ and the unknown spectrum $\mathbf{x}$. Bottom, a specific analyte concentration is estimated from the nth row of $\mathbf{S}^+$ and the spectrum $\mathbf{x}$

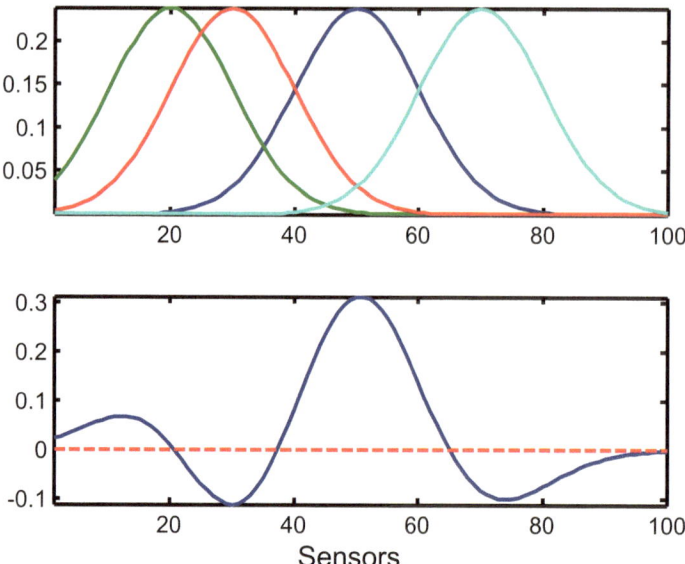

Fig. 2.6 (**a**) Spectra of four pure constituents in a range of 100 wavelengths or sensors. The blue line corresponds to the analyte of interest. (**b**) Vector of CLS regression coefficients associated to the analyte of interest

As an example, Fig. 2.6a shows the spectra of four pure constituents of an analytical system, defined in a range of 100 wavelengths. A typical calibration of this system using the CLS model will yield the vector of regression coefficients shown in Fig. 2.6b for the constituent with the blue spectrum in Fig. 2.6a. As can be seen, the vector $\mathbf{b}_n$ for this analyte shows positive and negative features. The most intense positive portion agrees with the position of the maximum of the pure spectrum of constituent n (identified by the blue spectrum), meaning that $\mathbf{b}_n$ carries a certain qualitative information on the analyte.

The vector of regression coefficients plays a prime role in multivariate calibration. Because $\mathbf{b}_n$ has the same number of elements as the number of spectral wavelengths, it is usual to speak about the *spectrum* of regression coefficients. This spectrum is quite abstract, with positive and negative portions, and thus it does not represent a real constituent. However, it does possess a certain qualitative value: the analyst expects that at least some intense bands of the $\mathbf{b}_n$ spectrum will correspond to real absorption bands of the analyte of interest. In the future we will return to these subjects in several cases, because analogous $\mathbf{b}_n$ vectors will appear in all multivariate models, although each of them will be estimated in a different manner from the calibration signals and concentrations.

Some companies developing NIR calibrations for industrial analytical purposes call the set of elements of the $\mathbf{b}_n$ vector *the equation*, and advertise it as an industrial product. If the reader purchases a NIR spectrometer and pays the company for developing calibrations, they will sell him the equation, which are in fact the elements of the vector of model regression coefficients for the analyte of interest. The instrument software will estimate the analyte concentration in an unknown sample through Eq. (2.14), from the elements of *the equation* (b_{jn}) and the signal values (x_j).

2.7   A CLS Algorithm

An algorithm consists of a series of instructions, written in a computer language, which will start from certain initial conditions, and will be executed sequentially until reaching the end. The most employed language in chemometrics is MATLAB,[1] a highly efficient environment in which a large number of models and procedures have been programmed, most of them freely available on the internet.

A CLS model can be easily programmed in MATLAB (see Box 2.1).

Box 2.1

After loading in the working space the variables "X" (the matrix of calibration signals), "Y" (the matrix of calibration concentrations), and "x" (the vector of signals for the unknown sample), two program lines are enough to estimate the matrix "S" and the constituent concentrations in the unknown "y":

```
S=X*Y*inv(Y'*Y);
y=inv(S'*S)*S'*x;
```

Notice that the size of the input variables should be: "X," $J \times I$ (J = number of wavelengths or sensors, I = number of calibration samples); "Y," $I \times N$ (N = number of analytes to be calibrated); "x," $J \times 1$. Those generated during program execution are "S," $J \times N$ and "y," $N \times 1$.

2.8   Validation Phase

Once the CLS model is calibrated, it is important to validate it. For this purpose, it is usual to prepare an independent validation sample set, in which the constituents are present at concentrations which are different than those employed for calibration. The validation concentrations are usually selected as random numbers within the corresponding ranges for each constituent. The comparison of the analyte concentrations estimated for the validation set is made using appropriate statistical indicators. One of them is the root mean square error in prediction (RMSEP), calculated according to:

---

[1]MATLAB. The Mathworks, Natick, Massachusetts, USA.

$$\text{RMSEP} = \sqrt{\frac{\sum\limits_{n=1}^{N_{\text{val}}} \left(y_{\text{nom},n} - y_{\text{pred},n}\right)^2}{N_{\text{val}}}} \tag{2.15}$$

where N_{val} is the number of validation samples, $y_{\text{nom},n}$ is the nominal concentration of analyte n in the validation samples, and $y_{\text{pred},n}$ is the CLS predicted concentration in the same samples.

Another common indicator is the relative error of prediction (REP), given in %, and defined as:

$$\text{REP} = 100\frac{\text{RMSE}}{<y>} \tag{2.16}$$

where $<y>$ is the mean value of the calibration concentrations for the analyte.

Incidentally, it is not uncommon to see published works where the RMSEP and REP are reported with unreasonably large numbers of significant figures; a parameter indicating uncertainty, and all parameters derived from uncertainty, should be reported with a single significant figure, or at most two, which is justified when the first figure is 1, or the first is 1 and the second is less than 5 (Olivieri 2015).

There are more sophisticated tests for the comparison of nominal and predicted concentrations. Beyond the statistical importance of these indicators, the visual impression of the analyst is most important. In this sense, it is usual to plot the predicted vs. nominal analyte concentration, and the prediction errors (difference between predicted and nominal) vs. predicted values. Figure 2.7a, for example, is reassuring: the relationship between predicted and nominal is reasonable, and approaches the ideal line of unit slope (Fig. 2.7a), with prediction errors showing a seemingly random distribution (Fig. 2.7b).

Figure 2.8, on the other hand, shows a different scenario, with reasonable results for most validation samples, except for one of them, whose predicted value and

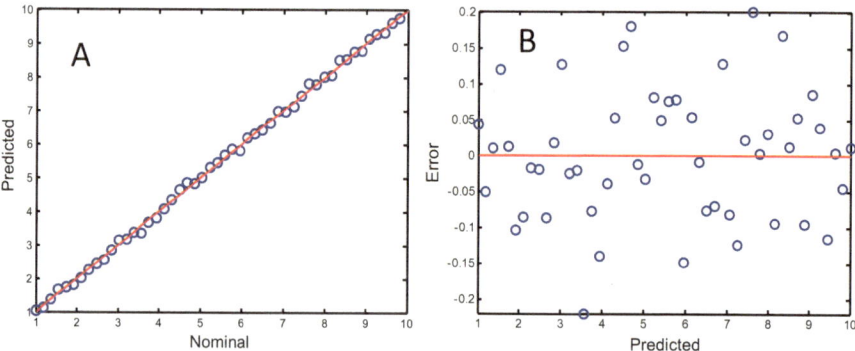

Fig. 2.7 (a) Predicted concentrations as a function of nominal values for the analyte of interest in 50 validation samples. (b) Prediction errors as a function of predicted concentrations. In (a) the red line indicates perfect correlation with unit slope; in (b) the red line indicates null errors

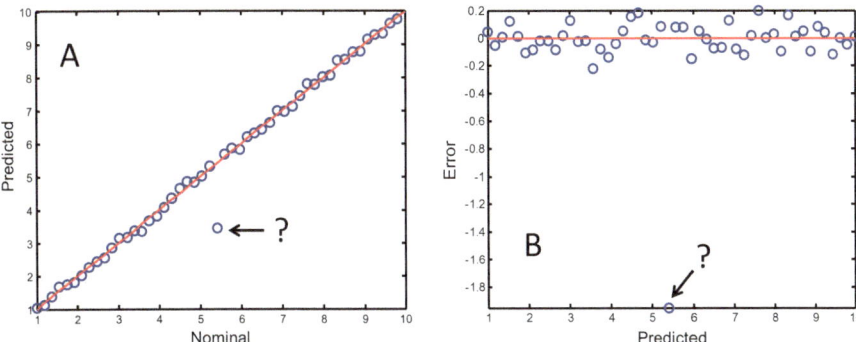

Fig. 2.8 (a) Predicted concentrations as a function of nominal values for the analyte of interest in 50 validation samples. One of the samples shows a significant deviation. (b) Prediction errors as a function of predicted concentrations. In (a) the red line indicates perfect correlation with unit slope; in (b) the red line indicates null errors. The problematic sample is indicated with a question mark

prediction error considerably deviate from the expectations. What to do with samples of this kind? The temptation of discarding them from the analysis may be irresistible. However, the analyst should go beyond this feeling, asking himself what really happened with this particular sample. Two main reasons may be responsible for this strange behavior: (1) preparation errors, making the nominal analyte concentration different than the one considered in Fig. 2.8 and (2) presence of interferent agents, not taken into account in the calibration phase, so that the prediction is biased. In any case, it is worth investigating these possibilities, repeating the preparation of the sample to verify reason (1), or using the diagnostic discussed in the next section to verify reason (2).

2.9   Spectral Residuals and Sample Diagnostic

In the framework of the CLS model, one may obtain a typical parameter of all least-squares fitting procedures: the regression residues. In the present case they are the elements of the vector $\mathbf{e}$ in Eq. (2.8), containing the uncertainty associated to the model of the unknown sample signal. The vector of residues can be found from Eq. (2.8) and the calibration matrix $\mathbf{S}$:

$$\mathbf{e} = \mathbf{x} - \mathbf{S}\,\mathbf{y} \tag{2.17}$$

It is also important to estimate, for each unknown sample, the standard deviation of the residuals s_{res}:

$$s_{\mathrm{res}} = \sqrt{\frac{\sum_{j=1}^{J} \left(e_j\right)^2}{J - N}} \tag{2.18}$$

Notice the $(J - N)$ degrees of freedom in Eq. (2.18), due to the fact that the sample provides J data (the signals measured at J wavelengths), and N parameters are estimated (the concentrations of the N analytes in the sample).

The value of s_{res} is indicative of the fitting success. As in all least-squares fitting procedures, one expects that the lack of fit (LOF) is not significant. The LOF is estimated through the comparison of s_{res} with the level of instrumental noise in the spectral measurements. The rigorous statistical comparison is made through an F-test between the square of s_{res} and the squared noise level (Miller and Miller 2010), as will be discussed in detail in a future chapter for advanced multivariate models. Intuitively speaking, the analyst expects that the value of s_{res} is of the order of the instrumental noise; one has always certain knowledge about the latter from the experience in using the instrument.

Could a CLS model show a significant lack of fit? If so, when? This could occur if Eq. (2.8) is not properly modeling the signals for the unknown. The main cause of a significant LOF lies in the presence of new constituents in the unknown sample, which may produce signals in the same spectral range of the calibration constituents. In the next section we will study this case in detail, but we may anticipate here a rather bitter-sweet result: the model will not work, in the sense that it will not be useful for predicting the analyte concentration in samples with significant LOF, but will warn the analyst about this fact.[2]

2.10   The First-Order Advantage

In the framework of the CLS model, it is assumed that the unknown samples do not possess constituents which are not present in the calibration sample set, and are able to produce a signal overlapping with those for the analytes. We now study the case of a new sample composed of substances not taken into account in the calibration phase. The result is that a significant error will occur in the prediction phase, mainly because Eq. (2.8) would not be correct. In the latter expression, only calibrated analytes should be present in the unknown sample.

This outcome is similar to that found in univariate calibration, where complete selectivity towards the analyte of interest is required for accurate prediction. However, in CLS calibration, the model can warn the analyst on the presence of uncalibrated phenomena in a given unknown sample. In this case, the value of the residual standard deviation s_{res} will be abnormally high in comparison with the instrumental noise level. Moreover, the visual inspection of the elements of the vector $\mathbf{e}$ in Eq. (2.8) may provide additional information. Since $\mathbf{e}$ has a number of elements

---

[2]One could also estimate the calibration spectral residues, which are the elements of the matrix $\mathbf{E}$ in Eq. (2.3), as $\mathbf{E} = \mathbf{X} - \mathbf{SY}^{\mathrm{T}}$. An analogous expression to Eq. (2.18), applied to the values of e_{ji} (the elements of $\mathbf{E}$), would measure the calibration fit. The LOF value for the calibration CLS phase is expected to be significant if: (1) the signal–concentration relationship deviates from the linear law, (2) important interactions among constituents occur, or (3) wrong constituent concentrations are introduced in matrix $\mathbf{Y}$.

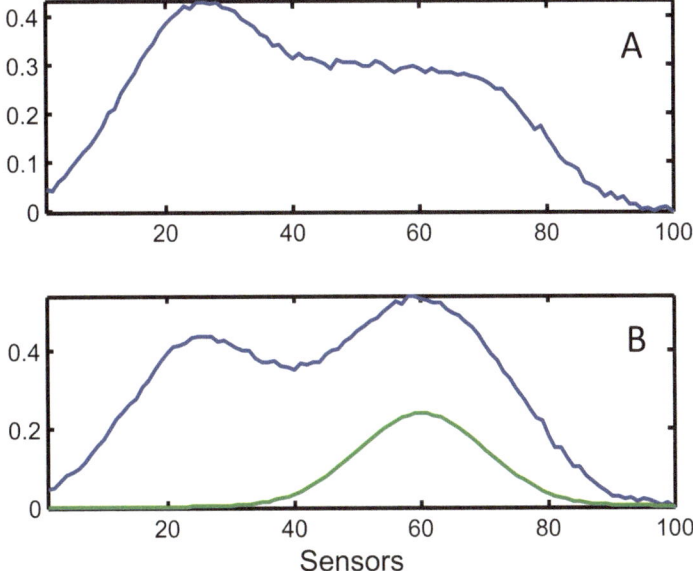

Fig. 2.9 (a) Spectrum of an unknown sample containing four constituents, whose spectra are shown in Fig. 2.6a. (b) Spectrum of an unknown sample (blue line) containing the same four constituents and an additional one whose pure spectrum is shown as a green line

equal to the number of wavelengths, a plot of **e** is called the spectrum of residuals. This spectrum will show random noise at all sensors if the fit to Eq. (2.8) is reasonably good, and the LOF is not significant. Conversely, if the LOF is significant, meaning that Eq. (2.8) does not adequately represent the chemical composition of the unknown sample, the spectrum of residuals will show positive and negative features of significant intensity.

In this way, multivariate models such as CLS are able to provide the analyst with information about the presence of unmodeled constituents in unknown samples. They may not be able to compensate for the presence of these constituents, but at least they can warn the analyst on these anomalies. This property is called the *first-order advantage*. Figures 2.9 and 2.10 illustrate these observations. Figure 2.9 shows the spectra of two unknown samples; in part a the sample is composed of the same four constituents of Fig. 2.6a, whereas in part b the sample contains, in addition to these four constituents, an unmodeled compound whose spectrum shows a maximum centered at sensor No. 60. If we submit both of these samples to a CLS model built with the four constituents of Fig. 2.6a, the resulting spectral residuals from Eq. (2.17) would be of the kind shown in Fig. 2.10. In part a of this latter figure, the spectrum of residuals is random and of low intensity, corresponding to a test sample which is representative of the calibration composition. In this case, s_{res} would be small and compatible with the level of instrumental noise present in the spectral measurements. In part b, on the other hand, the residuals are significantly higher in intensity than in part a, and suggest a spectral shape which differs from random

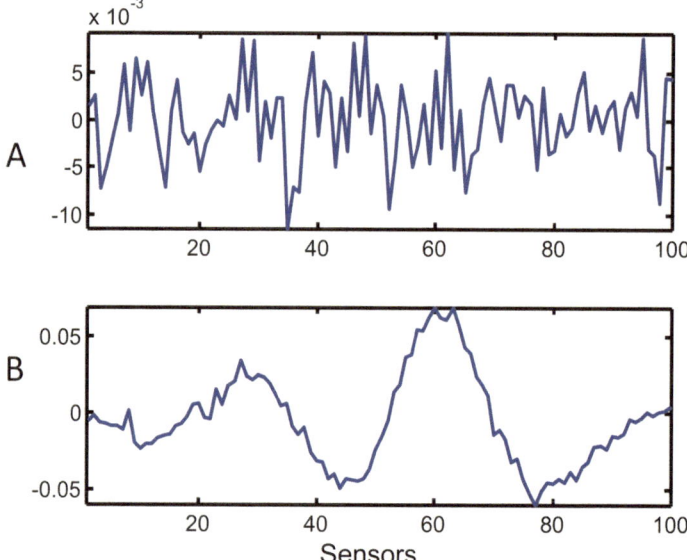

Fig. 2.10 (**a**) Spectrum of spectral residues after CLS analysis of the unknown sample of Fig. 2.9a. (**b**) The residues after CLS analysis of the unknown sample of Fig. 2.9b

noise. Moreover, the spectrum of residuals in part b shows a positive signal at the same sensor corresponding to the maximum for the interferent (sensor No. 60). In some way, the spectrum of residuals is indicative about which spectral regions correspond to the unexpected interferents. They do not provide their real spectra, but only qualitative indications that something odd is going on in a certain spectral region.

2.11   A Real Case

In this section we present a literature work where CLS was applied to the resolution of a real case. It deals with the simultaneous determination of three active principles in a pharmaceutical form: the antibiotics rifampicin, isoniazid, and pyrazinamide (Goicoechea and Olivieri 1999). Figure 2.11 shows the UV-visible absorption spectra of pure standard of the active principles. As can be seen, it is not possible to find a wavelength at which one analyte is the only responsive constituent (except rifampicin at wavelengths longer than 400 nm), which is the requirement for the successful application of classical univariate calibration. In pharmaceutical quality control, this type of determinations is usually carried out by means of liquid chromatography with UV-visible detection, so that the lack of selectivity of the spectroscopic technique is compensated by the selectivity of the chromatographic column, which physically separates the sample constituents before detection.

Fig. 2.11 UV-visible absorption spectra of the analytes of a real sample: rifampicin (green line, 6.80×10^{-6} mol L^{-1}), isoniazid (black line, 2.00×10^{-5} mol L^{-1}), and pyrazinamide (red line, 1.20×10^{-4} mol L^{-1})

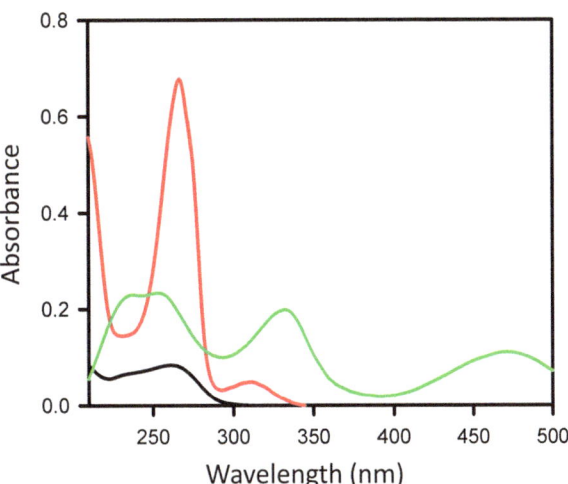

The work was required by a pharmaceutical company, due to the advantages of the spectral analysis in comparison with chromatography. We could mention the following ones: (1) it is faster, because a UV-visible spectrum may take a few seconds to be recorded, in contrast to chromatographic runs, with regular times in the order of minutes, (2) it does not employ auxiliary reagents or organic solvents, (3) it may be implemented with portable instruments outside the laboratory, (4) instruments may be connected remotely with optical fibers to the system, allowing to take informed decisions, etc.

It is important to notice that commercial samples of these pharmaceutical forms contain additional constituents to the active principles: the excipients. If any of these excipients dissolves when analyzing the tables, and absorbs UV-visible light in the working wavelength range, it will act as an interferent in the analysis, and should be included within the calibration constituents when planning the calibration phase. In the present experimental case, however, the excipients did not produce a significant interference, and the calibration set was designed with only three constituents: the active principles.

The calibration set included 15 mixtures of the three analytes. The concentrations of the analytes in these mixtures were in the range where Lambert–Beer's law was obeyed, and were established through an experimental design known as *central composite* (Leardi 2009). This design employs five different concentration levels for each analyte, as illustrated in the three-dimensional plot of the 15 concentrations (Fig. 2.12). As previously mentioned, the selection of the concentrations of calibration mixtures requires adequate designs providing representativity to the calibration, and minimizing the correlations among the columns of the matrix **Y**. Central composite designs put more emphasis in the combination of concentrations near the center of the design, in contrast to full-factorial designs, which include all possible combinations of concentration levels. This concept is best appreciated in Fig. 2.13, where a two-constituent central composite design is compared with a full-

Fig. 2.12 Three-dimensional representation of the concentrations of three analytes in a central composite design of 15 samples. The blue point is located in the center of the design, the eight black points are in the extremes of a factorial design at two levels, and the six red points (*star* points) outside the factorial design are located at a distance which depends on the number of constituents

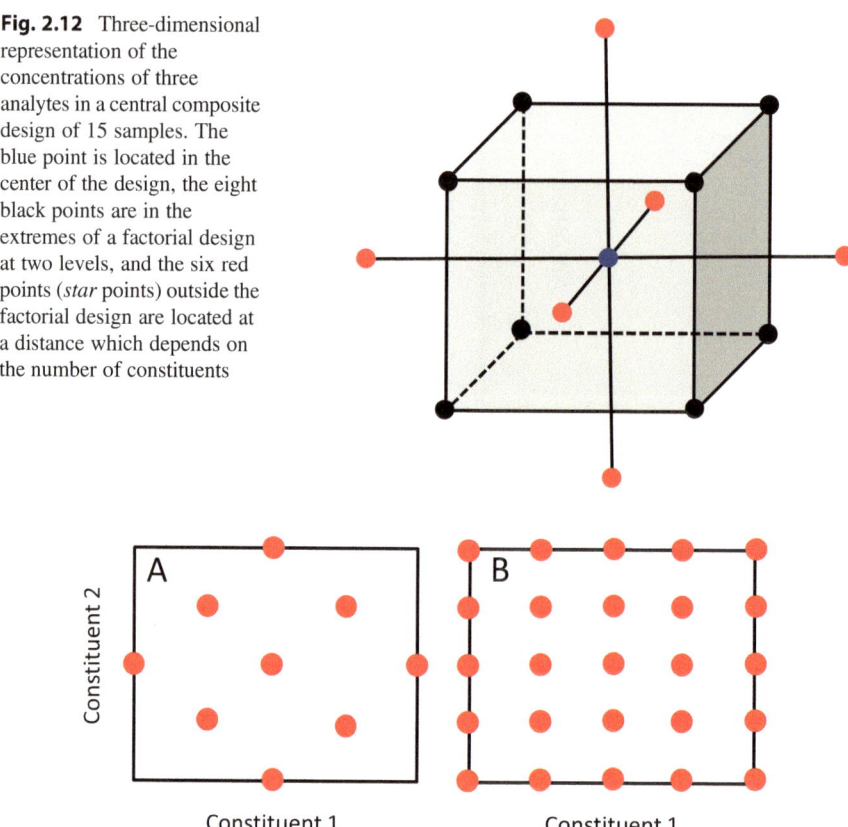

Fig. 2.13 Two possible mixture designs for a two-constituent analytical system with five different concentration levels. (**a**) Central composite. (**b**) Full factorial

factorial one. If both designs are built with five concentration levels, the central composite will require 9 experiments, whereas the full-factorial 25 (Fig. 2.13). The economy in experiments is apparent in the former case.

In the real experiment, once the spectra of the 15 calibration mixtures were registered, the CLS matrix **S** was found, and through its generalized inverse, the concentrations of the three analytes were predicted in a set of 6 validation samples, prepared with standards of the same three active principles. The predicted concentrations were compared to the nominal ones in the validation samples, and the result was … not good! According to the authors of this work, … *while the concentrations of rifampicin and pyrazinamide could be reasonably determined, the prediction errors for the less favorable analyte, isoniazid, were poor and on the order of 50%* (Goicoechea and Olivieri 1999). Why? The reader should conclude, from the reading of this chapter, on the reason of the poor predictive ability of the CLS model in this case, particularly with respect to the analyte isoniazid (Fig. 2.11).

We will discuss this system again in future chapters, by resorting to other multivariate calibration models.

2.12   Advantages and Limitations of CLS

The main advantages of CLS can be summarized as follows. It is based on a simple mathematical model, which can be easily implemented with the help of standard matrix calculations. If the samples to be analyzed do not present serious interferents, and no significant spectral collinearities are found among the analyte spectra, CLS analysis provides a rapid, simple, and reliable way of simultaneously estimating all analyte concentrations in multi-component samples.

The disadvantages are easy to imagine: it is sensitive to spectral correlations or collinearities, so that analytes with severely overlapped spectra cannot be reliably determined using CLS. It is also necessary to know all the chemical constituents present in the unknown samples, otherwise the presence of non-modeled interferents will lead to serious errors in the determination. This latter condition is hard to be fulfilled in complex samples of natural, biological, industrial, or environmental origin. The application range of CLS is thus extremely limited.

2.13   Exercises

1. Table 2.1 shows three different alternatives for the concentrations of two analytes in two calibration samples, which would be used to build a CLS model.
   (a) Write the matrix $\mathbf{Y}$ and the matrix product $(\mathbf{Y}^T\mathbf{Y})$ for each alternative
   (b) Calculate the determinant of the $(\mathbf{Y}^T\mathbf{Y})$ matrix in each case. Hint: for a 2×2 matrix $\mathbf{M}$, the determinant is simply $(m_{11}\, m_{22} - m_{12}\, m_{21})$
   (c) Which alternative is better for calibrating the CLS model?

Table 2.1   Three alternatives for the calibration concentrations of two analytes

Alternative 1		
Analyte	Sample 1	Sample 2
1	1	0
2	0	1
Alternative 2		
Analyte	Sample 1	Sample 2
1	1	1
2	2	2
Alternative 3		
Analyte	Sample 1	Sample 2
1	1	2
2	2	1

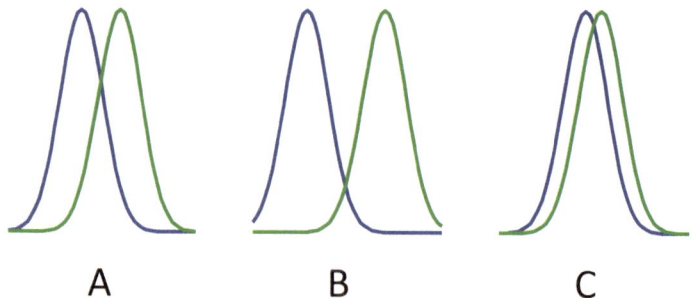

Fig. 2.14 Three alternative overlapping situations between the spectra of two pure analytes

2. Indicate in which of the following cases a CLS model would be successful and justify the answer.
   (a) Determination of the quality of chicken meat by NIR spectroscopy
   (b) Determination of the concentrations of two chemical constituents in the presence of a constant background signal by UV-visible spectroscopy
   (c) Determination of the concentrations of two chemical constituents in the presence of a variable background signal by UV-visible spectroscopy
3. Figure 2.14 shows three cases of different spectral overlapping between two analytes in pure form.
   (a) Order the three overlapping situations according to decreasing values of the determinant of the matrix $(\mathbf{S}^T\mathbf{S})$
   (b) Which situation is the most favorable one for predicting the analyte concentrations in an unknown sample?

References

Goicoechea, H.C., Olivieri, A.C.: Simultaneous determination of phenobarbital and phenytoin in tablet preparations by multivariate spectrophotometric calibration. Talanta. **47**, 103–108 (1998)

Goicoechea, H.C., Olivieri, A.C.: Simultaneous determination of rifampicin, isoniazid and pyrazinamide in tablet preparations by multivariate spectrophotometric calibration. J. Pharm. Biomed. Anal. **20**, 681–686 (1999)

Leardi, R.: Experimental design in chemistry: a tutorial. Anal. Chim. Acta. **652**, 161–172 (2009)

Miller, J.N., Miller, J.C.: Statistics and Chemometrics for Analytical Chemistry, 6th edn. Prentice Hall, New York (2010)

Olivieri, A.C.: Practical guidelines for reporting results in single- and multi-component analytical calibration: a tutorial. Anal. Chim. Acta. **868**, 10–22 (2015)

Tellinghuisen, J.: Inverse vs. classical calibration for small data sets. Fresenius J. Anal. Chem. **368**, 585–588 (2000)

Thomas, E.V., Haaland, D.M.: Partial least-squares methods for spectral analyses. 1. Relation to other quantitative calibration methods and the extraction of qualitative information. Anal. Chem. **60**, 1193–1202 (1988)

The Inverse Least-Squares Model

3

Abstract

The first and simplest inverse least-squares calibration model, also called multiple linear regression, is discussed in detail. Advantages and disadvantages are discussed for a model which today is still in use for some applications. Proposals are given for developing advanced calibration models.

3.1   Why Calibrating Backwards? A Brilliant Idea

The name *inverse calibration* is due to the use of the linear response-concentration law written in an inverse manner, in comparison with classical direct methods such as CLS. As will be shown below, inverse methods allow one to study multi-component samples where only one or a few analytes are of interest, but the concentrations, spectra, and chemical identities of the remaining sample constituents are in general unknown. In this way, they allow one to overcome the main disadvantage of CLS: the need of knowing the concentrations of all possible sample constituents in the calibration phase.

The development of inverse models began in the decade of 1960, because of the need of applying NIR spectroscopy to the analysis of materials. It was clear that NIR data had a great potential for the non-invasive analysis of intact material, particularly for the detection of quality parameters of seeds and other agricultural products. The main drawback was the high degree of spectral overlapping among the spectra for the analytes of interest (in fact, some of them are not really analytes but global properties) and the remaining sample constituents. The CLS model is not feasible in these cases, so that researchers focused on alternative mathematical models. The first work describing a new approach to this problem was made by Karl Norris, in the US Department of Agriculture, correlating the measurement of NIR data at two wavelengths with the moisture level in seeds (Norris and Hart 1965). This work

© Springer Nature Switzerland AG 2018   39
A. C. Olivieri, *Introduction to Multivariate Calibration*,
https://doi.org/10.1007/978-3-319-97097-4_3

paved the way to seemingly more ambitious aims, e.g., the use of whole NIR spectra in diffuse reflectance mode, for the direct analysis of samples without any previous treatment. The publication consolidating this approach is the most cited one in the field of NIR spectroscopy (Ben-Gera and Norris 1968). It is important to notice that Norris was not a mathematician, a statistician, or a chemometrician, but an agricultural engineer. This is perhaps an important lesson to be learned from the history of multivariate calibration: you do not need to be a mathematician to produce a mathematical revolution in analytical chemistry.

3.2   Calibration Phase

We now explore the first versatile method for the analysis of multi-component mixtures: inverse least-squares regression (ILS), a model also known as multiple linear regression (MLR). As previously discussed for the CLS model, direct calibration implies measuring the spectra of calibration samples containing the analytes at known concentrations, and obtainment of the matrix of sensitivities from the direct law by least-squares fitting of the following representative expression:

$$\text{Signal} = \text{Sensitivity} \times \text{Concentration} + \text{Errors} \tag{3.1}$$

In inverse calibration, on the other hand, the linear law is written inversely:

$$\text{Concentration} = \text{Signal} \times \text{Regression coefficient} + \text{Errors} \tag{3.2}$$

where proportionality is assumed between the concentrations of the calibrated constituents and the corresponding instrumental responses, through a set of regression coefficients to be estimated. Notice that the term collecting the model errors in Eqs. (3.1) and (3.2) are different: in the former case they correspond to spectral errors; in the latter, to concentration errors.

In inverse models, Eq. (3.2) is applied when the analyst only knows the concentration of a single analyte (or a few) in the calibration samples, but knows nothing about the remaining sample constituents. This highly important concept is the basis of the most powerful multivariate calibration models. The literature on the subject is abundant; we recommend the classical text of Massart et al. (1997), and the excellent article by Haaland and Thomas (1988).

A general model based on the inverse of Lambert–Beer's law can be expressed as:

$$\mathbf{Y} = \mathbf{X}^\mathsf{T}\mathbf{B} + \mathbf{E} \tag{3.3}$$

where the matrix $\mathbf{X}$ (of size $J \times I$) collects the instrumental signals for I calibration samples, measured at J wavelengths. The matrix $\mathbf{Y}$, on the other hand, contains the calibration concentrations, in each of the I samples, of each of N sample constituents, and its size is $I \times N$ (N is the total number of constituents). Up to here, both $\mathbf{X}$ and $\mathbf{Y}$ have identical meaning as in the CLS model, since $\mathbf{Y}$ contains the concentrations of all sample constituents.

In Eq. (3.3), **B** is a matrix of size $J \times N$ relating the concentrations with the responses in an inverse manner, and is called the matrix of regression coefficients. It contains all the regression coefficients relating each of the constituent concentrations with the calibration spectra. Finally, **E** is a matrix of error models, of size identical to **Y**. The presence of **E** in Eq. (3.3) is due to the same reasons why a similar error matrix was considered in the CLS calibration model. If the problem is over-determined, with more equations than unknowns, the estimation of the parameters is made by least-squares fitting, minimizing the sum of the squared elements of **E**.

Let us suppose now that we are not interested in all chemical constituents of the samples, but just on a few, possibly on a single one. It is likely that we know nothing about those other constituents, and only a single analyte is of interest. In this case, we do not need to know all the elements of the matrix **B**, but only the column corresponding to the analyte of interest, identified by the index n. Thus it is possible to propose a simplified model in which only the concentrations for analyte n are known, isolating from Eq. (3.3) the portion corresponding to the latter analyte:

$$\mathbf{y}_n = \mathbf{X}^T \mathbf{b}_n + \mathbf{e} \qquad (3.4)$$

where $\mathbf{y}_n$ is a vector of analyte concentrations n, $\mathbf{b}_n$ is the column of the matrix **B** corresponding to this analyte, and **e** is the vector of concentration residuals for this reduced ILS model. Figure 3.1 shows by means of blocks the meaning of Eqs. (3.3) and (3.4).

Formally, finding $\mathbf{b}_n$ from Eq. (3.4) implies the same steps as in the CLS model: pre-multiplication of both sides by **X**, followed by pre-multiplication by the inverse $(\mathbf{X}\,\mathbf{X}^T)^{-1}$:

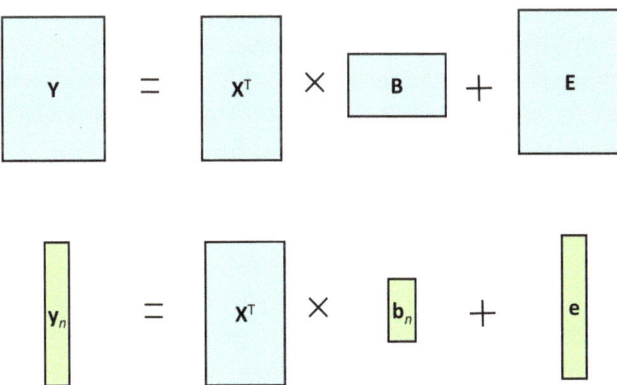

Fig. 3.1 Top, block-based scheme corresponding to Eq. (3.3), representing the ILS model for all the chemical constituents of the calibration samples, whose concentrations are grouped in the matrix **Y**. The matrix **X** contains the spectra of the calibration samples in its columns, the matrix **B** the vectors of regression coefficients in its columns, and the matrix **E** collects the model errors. Bottom, isolation of the portion of the top scheme corresponding to a single analyte of interest: a specific column of **Y** and a specific column of **B** are isolated and called $\mathbf{y}_n$ and $\mathbf{b}_n$, respectively

$$\mathbf{b}_n = (\mathbf{X}\mathbf{X}^{\mathrm{T}})^{-1}\mathbf{X}\mathbf{y}_n \tag{3.5}$$

Defining the generalized inverse of $\mathbf{X}$ as $\mathbf{X}^+$ allows one to condense Eq. (3.5):

$$\mathbf{b}_n = \mathbf{X}^+\mathbf{y}_n \tag{3.6}$$

Equation (3.6) illustrates the main difference between ILS, where only the analyte concentrations need to be known, and CLS, where all concentrations of all constituents should be known in the calibration samples. It is important to recall that the matrix $\mathbf{X}$ should always be representative of the composition of future unknown samples.

Notice that Eq. (3.6) does not mean that the ILS model allows for the study of a single analyte only. If there are more analytes of interest, one could in principle write a separate model for each of them, analogous to Eq. (3.6). For each analyte, a specific vector of calibration concentrations $\mathbf{y}_n$ is needed, leading to an analyte-specific regression vector $\mathbf{b}_n$.

Finally, the ILS model allows one to study global sample properties (organoleptic properties of food, octane number in gasoline, etc.). In Eq. (3.6), the vector $\mathbf{y}_n$ would not contain specific analyte concentrations, but global sample properties. ILS does not appear to face major issues for the calibration phase. However, a detailed analysis of the mathematical requirements involved might bring some bad news.

3.3   Mathematical Requirements

Equation (3.4) is the formal expression for a system of simultaneous equations with multiple unknowns. One such equation is shown in Fig. 3.2, where the concentration of the analyte of interest in sample 1 (y_1) is expressed as the sum of products of the signal at each wavelength for this particular sample (x_{11}, x_{21}, ...) and the corresponding regression coefficient (b_1, b_2, ...).

The analysis in terms of number of equations and unknowns is as follows. There are I equations (the number of elements of $\mathbf{y}_n$, which is the number of calibration samples). On the other hand, J unknowns should be estimated, i.e., the elements of the vector of regression coefficients $\mathbf{b}_n$, a number which is coincident with the number of spectral wavelengths (or sensors in other multivariate signals). The problem will be over-determined, in principle, if $I > J$. We face here the first problem when trying to calibrate an ILS model: it requires more calibration samples than wavelengths.

One option is to prepare a large number of calibration samples, generously larger than the number of wavelengths. This might be demanding in terms of cost or time, and may require several thousands of samples. Another alternative is to select, from the whole spectral range, a reduced group of wavelengths, so that the samples outnumber the wavelengths. This requires a convenient algorithm to select the working wavelengths, and may lead to a significant loss of sensitivity, by discarding sensors which are potentially useful for the analysis.

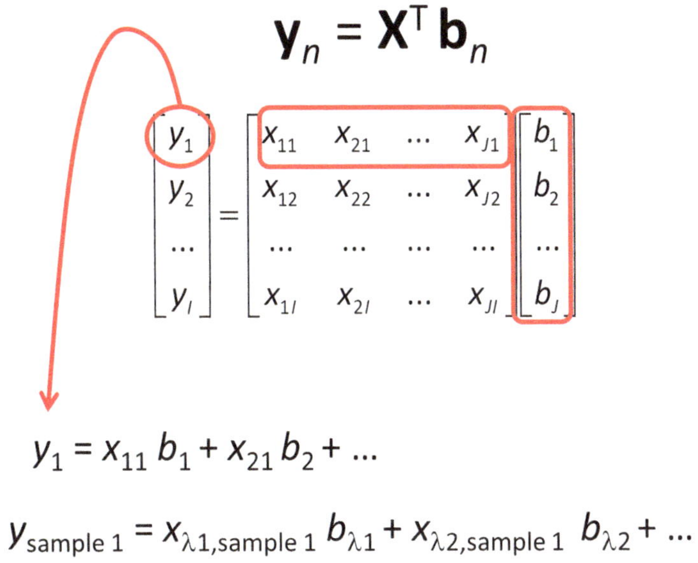

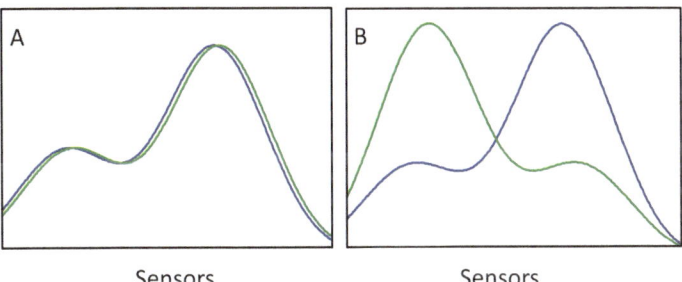

$$\mathbf{y}_n = \mathbf{X}^T \mathbf{b}_n$$

$$\begin{bmatrix} y_1 \\ y_2 \\ \dots \\ y_I \end{bmatrix} = \begin{bmatrix} x_{11} & x_{21} & \dots & x_{J1} \\ x_{12} & x_{22} & \dots & x_{J2} \\ \dots & \dots & \dots & \dots \\ x_{1I} & x_{2I} & \dots & x_{JI} \end{bmatrix} \begin{bmatrix} b_1 \\ b_2 \\ \dots \\ b_J \end{bmatrix}$$

$$y_1 = x_{11}\, b_1 + x_{21}\, b_2 + \dots$$

$$y_{\text{sample 1}} = x_{\lambda 1,\text{sample 1}}\, b_{\lambda 1} + x_{\lambda 2,\text{sample 1}}\, b_{\lambda 2} + \dots$$

Fig. 3.2 A specific equation of the system of multiple equations in the ILS calibration phase. The concentration of the analyte in sample 1 is shown to be the product of the first row of $\mathbf{X}^T$ and the column $\mathbf{b}_n$

A · · · Sensors   B · · · Sensors

Fig. 3.3 (**a**) Two spectra with a high degree of overlapping. (**b**) Two spectra with a low degree of overlapping

This is not the end of the problems. Even if the condition $I > J$ is met, to safely continue the calibration process represented by Eq. (3.5), the matrix $(\mathbf{X}\mathbf{X}^T)$ should be inverted. This implies that the columns of $\mathbf{X}$, at the selected working wavelengths, should not show a significant degree of correlation or overlapping. Otherwise, the determinant of $(\mathbf{X}\mathbf{X}^T)$ will be zero, precluding the matrix inversion in Eq. (3.5), or close to zero, making the inversion unstable.

Figure 3.3 illustrates two extreme cases of spectral overlapping: in the first one (part A), the degree of overlapping is significant at all sensors. Spectra with these characteristics will lead to a very small value of the determinant of the matrix $(\mathbf{X}\mathbf{X}^T)$, independently on the specifically selected wavelengths. In the second one (part B),

Fig. 3.4 NIR spectra of a set of 50 corn seed samples, employed to build a model for the determination of quality parameters. They were measured in the wavelength range 1100–2498 nm at 2 nm intervals (700 channels), and are available at http://www. eigenvector.com/data/Corn

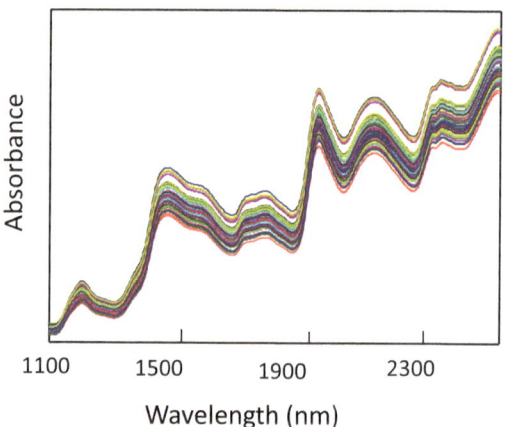

on the other hand, the overlapping is relatively low, although not null. If all columns of **X** are overlapped in this latter way, there will be no problems in inverting $(\mathbf{X}\,\mathbf{X}^{\mathrm{T}})$ at selected sensors. Unfortunately, in real-world applications using NIR spectroscopy, the calibration spectra resemble Fig. 3.3a rather than the pleasing Fig. 3.3b. See, for example, a set of NIR spectra for the calibration of typical parameters in corn seeds which is shown in Fig. 3.4. These calibration spectra display a high degree of overlapping or collinearity.

Selecting a reduced number of wavelengths with minimal correlation from a large number of available sensors is not a simple task, and various algorithms have been developed for this purpose, so that the ILS model can be successfully applied. In a future section, a variable selection algorithm will be described, which is still in use today. The degree of difficulties faced by the algorithm will be clear then.

3.4   Prediction Phase

During prediction, an inverse equation similar to the calibration phase is written [Eq. (3.3)], based on the inverse Lambert–Beer's law applied to the unknown sample:

$$y_n = \mathbf{b}_n^{\mathrm{T}}\mathbf{x} \tag{3.7}$$

where **x** is the spectrum of the unknown sample. Equation (3.7) implies that $\mathbf{b}_n$ behaves as the regression coefficient for constituent n (the *equation*), as previously discussed for CLS. If more than one analyte of interest occur, Eq. (3.7) is applied as many times as necessary, using each time the vector $\mathbf{b}_n$ associated to each analyte.

It is also important to mention that in the framework of ILS, no model errors are available for the test sample signals in the prediction phase [Eq. (3.7)]. Therefore, no information on possible unmodeled constituents in the unknown sample is provided, losing the important first-order advantage shown by CLS.

3.5 An ILS Algorithm

We now provide a set of MATLAB instructions for performing ILS (Box 3.1). The algorithm looks deceptively simple. However, notice that the variable "X" in Box 3.1 is not the full-spectral calibration data matrix, but the matrix of selected signals at suitable wavelengths. The task of selecting the specific wavelengths is considerably more complex, as we shall see below.

Box 3.1 ILS Algorithm

After loading in the workspace the variables "X," "yn," and "x," i.e., the calibration data matrix at the selected working wavelengths, the vector of analyte concentrations or sample properties in the calibration samples, and the spectral vector for the unknown sample (at the same wavelengths selected for calibration), it is simple to apply the following two MATLAB programming lines:

bn=inv(X*X')*X*yn;

y=x'*bn;

where "bn" is the vector of regression coefficients and "y" the predicted analyte concentration or sample property.

The size of the input variables should be: "X," $J \times I$ (J = number of wavelengths or sensors, I = number of calibration samples); "yn," $I \times 1$; and "x," $J \times 1$. Those generated during program execution are "bn," $J \times 1$ and "y," 1×1.

3.6 Validation Phase

The validation phase of the ILS model is analogous to the one described for CLS in the previous chapter. A group of new samples, independent of those employed for calibration, and whose analyte content or sample property is known, is submitted to the ILS model after measuring their spectra or multivariate signals. A comparison is then made of the prediction results with the nominal values using an appropriate statistical test. The statistical indicators are useful to judge on the analytical ability of the model with respect to these validation samples: root mean square error of prediction (RMSEP) and relative error of prediction (REP), as described in Chap. 2. In the future, the model can be monitored by means of new sets of samples, periodically analyzed using the developed multivariate ILS model and a reference technique to control the calibration maintenance.

The visual inspection of the plots of predicted vs. nominal concentration and prediction errors vs. predicted values will be useful, in the same way as described for CLS. However, if a sample is problematic (prediction error significantly larger than the average), with ILS it is not possible to detect the presence of interferents not modeled by the calibration phase. This is because ILS does not possess the first-order advantage, as discussed above.

3.7   Advantages and Limitations of ILS

The main advantage of the ILS model stems from the study of complex mixtures by an inverse calibration process, knowing only the concentrations of a limited number of constituents. In other words, ILS permits the quantitation of an analyte in the presence of interferences, provided the latter are properly included in the calibration phase, although their concentrations or chemical identities are not known.

The main disadvantage of ILS lies in the requirement of a limited number of working sensors, which leads to a loss of information and overall sensitivity, and to the need of selecting sensors with minimal correlation, which is not an easy task. Additionally, ILS does not show the important first-order advantage, already discussed for the CLS model. Due to these limitations, the modern practice has been gradually replacing ILS by more powerful methodologies, to be discussed in future chapters.

3.8   The Successive Projections Algorithm

The aim of this algorithm is to select a small number of variables (wavelengths or sensors) with a low degree of collinearity, so that the ILS model can be safely applied (Araújo et al. 2001; Soares et al. 2013). To gather an idea of the magnitude of the problem, imagine that 30 calibration samples are available, with spectra measured at 100 wavelengths. To build a successful ILS model, the number of wavelengths should be smaller than the number of samples, so that the question is: how may sub-sets of 1, 2, 3, etc. and up to 29 wavelengths can be built with these 100 wavelengths? Combinatorial theory tells us that the answer is 100 sub-sets of one wavelengths, $100 \times 99/2 = 4950$ sub-sets of two wavelengths, etc., making a total of 2×10^{25} sub-sets, i.e., a 2 followed by 25 zeros!

All of these models will provide a vector of regression coefficients $\mathbf{b}_n$, which will allow one to predict the analyte concentration in an independent group of samples. In this way, the models can be evaluated as regards their predictive ability, and the best one will be the one leading to the lowest prediction error. If each of these operations takes one microsecond, calculating all combinations would require 600,000,000,000 years! This shows the importance of developing intelligent algorithms for selecting the optimal sub-set of sensors, or a sub-set close to the optimum, allowing one to reach the goal rapidly and efficiently.

One of these methodologies is known as the successive projections algorithm (SPA). The idea is to produce a reduced number of sub-sets, each of them with minimal correlation among the selected variables, based on the concept of orthogonal (perpendicular) projection. SPA involves various phases, among which the first and most important one will be detailed here: the definition of the potential sub-sets of sensors. Additional procedures have been developed for improving the SPA results (Galvão et al. 2008).

In the above example, SPA will only produce a total of $100 \times 29 = 2900$ sub-sets, instead of 2×10^{25}. In a general case, the number of sub-sets will be $J \times (I - 1)$. The first 100 sub-sets of one wavelength are simple to define: each of them contains each of the 100 original variables at which the spectra were measured. The next 100 sub-sets contain two wavelengths. Instead of selecting all possible combinations of each of the 100 variables with each of the remaining 99 variables, SPA selects, as the second sensor accompanying the first one, the variable that best fulfills the requirement of perpendicularity (orthogonality) with respect to the first one.

The rationale behind SPA is the following: each sensor can be represented by a vector, or a set of I numbers: the row elements of the calibration matrix $\mathbf{X}$. These vectors point in a certain direction in a multi-dimensional sample space, and the tip of the vector indicates the position of the sensor. In our example, SPA applies the following protocol to find the 100 two-sensor sub-sets:

1. Consider the first one-sensor sub-set.
2. Find the projections of the remaining 99 variables in the direction perpendicular to the selected sensor in the sample space.
3. Select the variable with the largest orthogonal projection as the second sensor accompanying the first one.
4. Return to 1 and continue until the 100 sensors have been processed.

The above procedure will furnish 100 two-sensors sub-sets, instead of the 9900 of a comprehensive search.

The concept of maximum orthogonal projection is illustrated in Fig. 3.5. In the simplest possible sample space with just two samples, three variables are shown as arrows pointing in the directions of their corresponding rows of $\mathbf{X}$. If the first variable of a particular two-sensor sub-set is the blue one (λ_1), to find the second one, the projections of the remaining two variables (λ_2 and λ_3) are obtained in the direction perpendicular to λ_1. The largest is the green one (Fig. 3.5) so that λ_2 is selected as the second variable accompanying λ_1 in the two-sensor sub-set.

Fig. 3.5 Illustration of the process of successive orthogonal projections in the sample space. For a given sub-set of variables, λ_1 is the first selected sensor. The projections are then calculated of the remaining two variables λ_2 and λ_3 in the direction which is perpendicular to the first variable λ_1. The largest orthogonal projection is the one for λ_2, so that this latter sensor is selected as the second one in the sub-set starting with λ_1

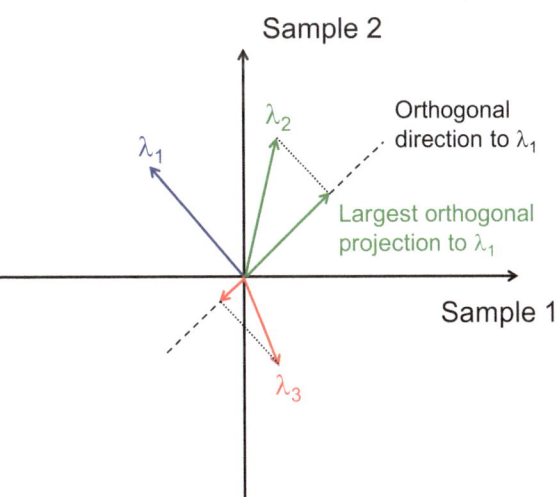

Continuing with three-sensor sub-sets in our example of 100 variables and 30 samples, a similar criterion is adopted: the third selected sensor accompanying each of the 100 two-sensor sub-sets will be the one, from the 98 remaining sensors, presenting the largest orthogonal projection with respect to a space defined by the first two variables. And so on, until reaching the last 100 sub-sets of 29 variables each. It is easy to understand the algorithm name: *successive projections*.

Once the 2900 sub-sets have been produced, each of them is probed by an ILS model to see which provides the best prediction results regarding the analyte concentration or sample property to be measured. Some procedures have been developed to efficiently filter the best variables of the best sub-set, leading to an even more reduced set of sensors with optimal analytical results (Galvão et al. 2008). A MATLAB graphical interface is freely available for the easy implementation of SPA (Paiva et al. 2012), which can be downloaded from http://www.ele.ita.br/~kawakami/spa/.

3.9   A Simulated Example

Algorithms such as SPA have renewed the interest in the use of ILS in multivariate calibration for solving complex analytical problems (Soares et al. 2013). A simulated example of the application of SPA is now discussed. Figure 3.6a shows the spectra of four constituents of an analytical system, in which the analyte of interest is the one with the blue spectrum, and Fig. 3.6b shows spectra for typical mixtures of the four constituents. If 30 four-constituent mixtures are prepared in random proportions to calibrate an ILS model, one should select, as discussed above, 2900 different sub-sets through SPA. After applying the latter algorithm, the best sub-set of sensors for building an ILS model is shown in Fig. 3.7. SPA only selects four wavelengths, located *by accident* near the maxima of the four known spectra of the pure constituents. In a general case, it is likely that the sample constituents or chemical nature of the analyte of interest are not known. SPA will then operate as a black box, just providing the optimal sensors or wavelengths for the analytical work.

The fact that four wavelengths were selected for building an ILS model in this simulated four-constituent example should not come as a surprise. Will SPA always select a number of wavelengths equal to the number of responsive components? In general no; the simulated case is too perfect, so to speak, the degree of spectral overlapping is rather low, there are no baseline effects, etc. In a general case, the real number of sample constituents is not known, and thus the relationship between SPA-selected wavelengths and constituent number is difficult to establish.

3.10   A Real Case

In this section a literature work will be discussed, in which the SPA method was applied to select NIR spectral wavelengths, with the aim of measuring some relevant parameters of fuel samples: sulfur content, distillation temperatures (starting point,

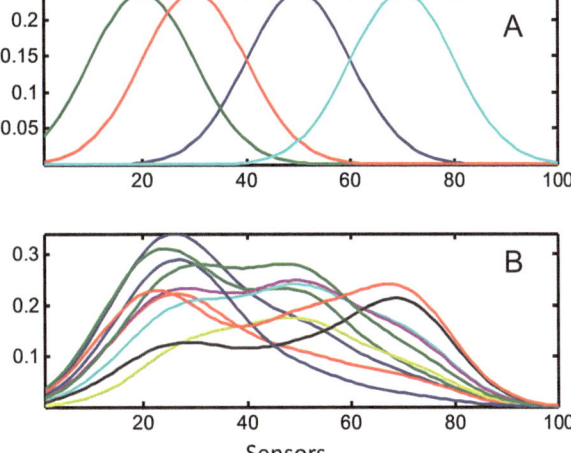

Fig. 3.6 (**a**) Spectra of four pure constituents. The blue spectrum corresponds to the analyte of interest. (**b**) Typical spectra of some calibration mixtures containing the four constituents

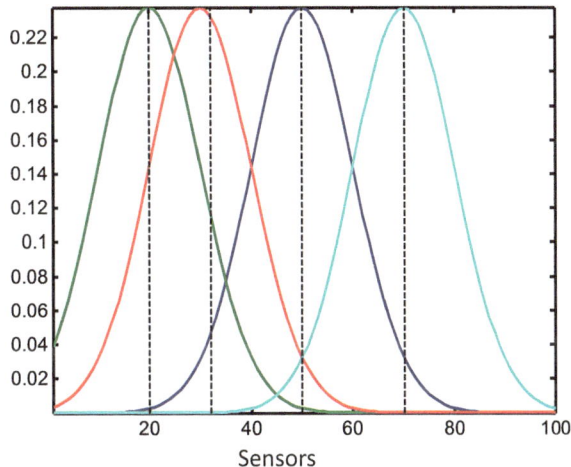

Fig. 3.7 Wavelengths selected by SPA (vertical dashed lines) to calibrate an ILS model for the analyte of interest (blue spectrum in Fig. 3.6)

and T10% and T90% parameters, temperatures at which 10% and 90% of the initial volume distill respectively) (Galvão et al. 2008). These parameters were measured according to the international ASTM (*American Society for Testing and Materials*) norms 4294-90 and D86. A set of 170 samples was available, whose spectra were collected at 1191 NIR wavelengths (Fig. 3.8). The sample set was divided in three sub-sets: one for calibration (65 samples), another one for optimizing the wavelength selection (40 samples), and a third one for independent validation (65 samples). In a first stage, the first derivative of all spectra was calculated, with the objective of reducing the effect of the dispersed NIR radiation in the analysis. The application of this and other mathematical pre-processing operations will be discussed in detail in a future chapter.

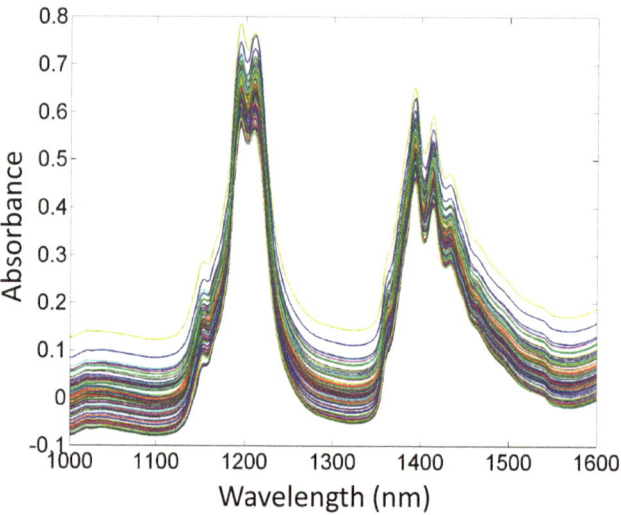

Fig. 3.8 NIR spectra of 170 diesel samples, employed to build an ILS model for calibration parameters of interest. Reproduced with permission from Galvão et al. (2008) (Elsevier)

The derivative spectra and the parameter values were used to select the optimal wavelengths for separate ILS calibrations. In the specific case of the parameter T10%, it is noticeable that from the 1191 original variables, SPA selected only 15, as detailed in Fig. 3.9. The ILS model built with these 15 variables led to a final prediction error of 3.5 °C in the T10% values, corresponding to a calibration range from 190 °C to 270 °C. Thus the relative error is only 1.5%. Predictions of a similar quality were obtained for the remaining parameters (Galvão et al. 2008).

In this real example, the value of the NIR technique and the ILS model can be appreciated for the calibration of global sample properties. As previously discussed, the analyst expects that the global sample properties depend on the chemical composition of the fuel, which in turn will be reflected in the NIR spectra.

3.11 How to Improve ILS: Ridge Regression

At this point we may ask what ILS needs to become a multivariate model with all the desired properties, without paying the price of selecting a few calibration wavelengths. Let us write a list of wishes:

1. Maintain the inverse calibration model, which guarantees calibration for a particular analyte (or a few analytes) or sample global properties, ignoring the chemical identities and concentrations of the remaining constituents.
2. Employ all available wavelengths or measuring sensors, ensuring maximum selectivity and sensitivity.
3. Permit the safe inversion of the calibration data matrix.

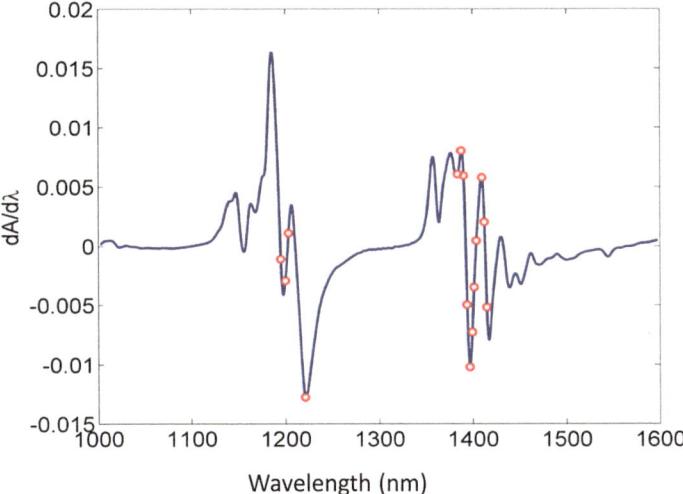

Fig. 3.9 Wavelengths selected by SPA (red circles) to calibrate an ILS model for the parameter T10% of diesel fuels, indicated on the first derivative of the average spectrum for the calibration samples. Reproduced with permission from Galvão et al. (2008) (Elsevier)

One alternative, still in use today by some researchers, is to solve the ILS Eq. (3.4) with all available wavelengths in the calibration matrix **X**, using a modified least-squares procedure known as *ridge regression* (RR) (Hoerl and Kennard 1970; Kalivas 2001). The specific aspects of the latter model are beyond the scope of this book, but a motivation for RR can be given here. In ILS, the least-squares solution intends to minimize the sum of squared errors for the model Eq. (3.4), so that the problem can be formally described as:

$$\mathbf{b}_n = \arg \min \left(\sum_{i=1}^{I} e_i^2 \right) \tag{3.8}$$

where e_i is an element of the residual concentration vector **e** in Eq. (3.4). Equation (3.8) is interpreted as saying that $\mathbf{b}_n$ is the argument leading to a minimum in the sum of squared calibration errors.

We know that full spectra lead to infinitely large regression coefficients in Eq. (3.8). In a future chapter, we will learn that large regression coefficients produce large prediction uncertainties, which are proportional to the length of the vector of regression coefficients $\left(\sqrt{\sum_{j=1}^{J} b_{jn}^2} \right)$, so that a trade-off can be set by minimizing a combination of squared calibration error and squared prediction error. The compromise is controlled by the RR parameter λ in the following way:

$$\mathbf{b}_n = \arg\min\left(\sum_{i=1}^{I} e_i^2 + \lambda \sum_{j=1}^{J} b_{jn}^2\right) \tag{3.9}$$

From Eq. (3.9), it can be shown that the vector of RR regression coefficients is given by:

$$\mathbf{b}_n = (\mathbf{XX}^{\mathrm{T}} + \lambda\mathbf{I})^{-1}\mathbf{Xy}_n \tag{3.10}$$

where $\mathbf{I}$ is an appropriately dimensioned identity matrix. What is important in the present context is that the matrix inversion in Eq. (3.10) is always possible, even when the matrix $(\mathbf{X}\,\mathbf{X}^{\mathrm{T}})$ is singular and cannot be inverted. Of course, RR requires one to optimize the value of λ, which can be done by different procedures (Golub et al. 1979).

We now show the performance of RR in a simulated case, based on the estimation of the vector of regression coefficients for the example of Fig. 3.6. Four constituents are present, with the pure spectra shown in Fig. 3.6a. If we assume that only the calibration concentration of the analyte of interest (blue trace in Fig. 3.6a) is known, and RR is applied with the optimized value of the parameter λ, the resulting vector of full-spectral regression coefficients is shown in Fig. 3.10a. Is this vector useful for future predictions in new samples? In this simulated situation, we can compare it with the one estimated by a CLS model from the calibration spectral matrix and the complete matrix of calibration concentrations for all four constituents (this matrix is known for this specific problem, but remains unknown in a general case, except for the column associated to the analyte of interest). The CLS vector of regression coefficients for the analyte of interest (see Chap. 2) is shown in Fig. 3.10b, and is almost identical to the one for RR. We may conclude that RR could be a viable alternative to improve the limitations of classical ILS.

A real application of RR can be cited: the determination of octane number in gasolines from near infrared spectroscopy (Chung et al. 2001). The authors found good prediction results, with average errors on the order of 0.3 units in the octane number, better than those furnished by ILS. Despite the viability of RR, the general consensus is that alternative models, to be discussed in future chapters, are preferable. The next section provides some insight into these advanced models.

3.12   An RR Algorithm

A short MATLAB algorithm for ridge regression is given in Box 3.2. Notice that its implementation requires to previously tune the parameter λ of Eq. (3.10).

Box 3.2 RR Algorithm
The ridge regression model is analogous to the ILS model, but the expression for the vector of regression coefficients slightly differs. A single MATLAB

(continued)

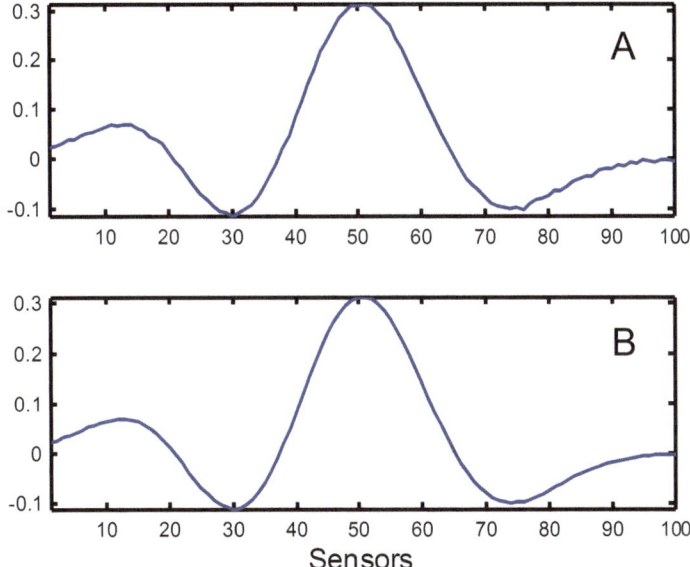

Fig. 3.10 Vectors of regression coefficients for the analyte of interest in the example of Fig. 3.6.
(**a**) From a ridge regression model, only knowing the vector of calibration concentrations for the
analyte of interest. (**b**) From a CLS model, knowing the matrix of calibration concentrations of all
four constituents

Box 3.2 (continued)
line is enough to estimate "bn," from the same workspace variables as in Box
3.1, except that the RR parameter "lambda" is required:
    bn=inv(X*X'+lambda*eye(size(X,1)))*X*yn;

3.13   How to Improve ILS: Compressed Models

Another alternative for improving the ILS model, and stemming from inspection of
the ILS calibration Eq. (3.4), is to replace the matrix of real signals $\mathbf{X}$ by another
matrix $\mathbf{T}$, with some specific properties. This matrix $\mathbf{T}$ should be a somehow
compressed version of $\mathbf{X}$, significantly smaller than $\mathbf{X}$ in size, to relax the requirement of a large number of samples. The compression process leading to $\mathbf{T}$ from
$\mathbf{X}$ should be able to retain the main properties of $\mathbf{X}$, in the sense of representing the
changes in spectra from sample to sample. Specifically, we need the following
inverse model:

$$\mathbf{y}_n = \mathbf{T}\,\mathbf{v}_n + \mathbf{e} \tag{3.11}$$

where $\mathbf{T}$ would play the role of $\mathbf{X}$, and $\mathbf{v}_n$ would involve a new set of regression coefficients adapted to $\mathbf{T}$. About 1960, compression methods capable of fulfilling these requirements were well known, leading to a rapid development of a model overcoming the problems of ILS. In fact, the specific employed compression method was known from the XIX century. The details will be given in the next chapters.

3.14   Exercises

1. Consider an analytical system composed of N chemical constituents, all known and available for the preparation of mixtures of standards.
   (a) Show that the matrix of ILS regression coefficients for all constituents ($\mathbf{B}$) is equivalent to the generalized inverse $\mathbf{S}^+$ of a CLS model
   (b) Show that in both the CLS and ILS models the number of wavelengths for calibration should be at least equal to N
   (c) Explain the results on wavelength selection by SPA in the system of Fig. 3.7.
2. In the real case described in Sect. 3.10, SPA selected 15 wavelengths for calibrating the parameter T10%. Does it mean that 15 chemical constituents have an influence on the target parameter?
3. In 1960, Norris found that the moisture level of a large number of wheat, soybean, wheat flour, and wheat bran samples could be reasonably predicted as a function of the difference of NIR absorbance at two wavelengths, 1940 and 2080 nm:

$$\text{moisture} = k(x_{1940} - x_{2080})$$

   (a) Write the latter expression in terms of an ILS model at two wavelengths
   (b) Give a specific expression for the vector of regression coefficients $\mathbf{b}_{\text{moisture}}$

4. Explain how an ILS model could be developed for calibration of the following systems:
   (a) Determination of the quality of chicken meat by NIR spectroscopy
   (b) Determination of the concentrations of two chemical constituents in the presence of a variable background signal by UV-visible spectroscopy
5. (a) Estimate the regression vector $\mathbf{v}_n$ from Eq. (3.11) by least-squares
   (b) Explain the mathematical requirements for the matrix $\mathbf{T}$ of Eq. (3.11) to be useful for developing a successful ILS model
6. There are literature precedents on the determination of metal ions in natural samples, such as the quantitation of calcium, potassium, magnesium, phosphorus, sodium, sulfur, iron, boron, and manganese in wine (Cozzolino et al. 2008). Given that simple metal ions do not show relevant vibrational NIR bands, how can you explain that metals can be quantitated by NIR spectroscopy and chemometrics?

7. Show that the vector of regression coefficients for ILS and for RR are those given above in this chapter, starting from the objective functions to be minimized, written in the following way:

$$f_{\text{ILS}} = \left(\sum_{i=1}^{I} e_i^2\right) = \mathbf{e}^{\text{T}}\mathbf{e} = (\mathbf{X}^{\text{T}}\mathbf{b}_n - \mathbf{y}_n)^{\text{T}}(\mathbf{X}^{\text{T}}\mathbf{b}_n - \mathbf{y}_n)$$

$$f_{\text{RR}} = \left(\sum_{i=1}^{I} e_i^2 + \lambda \sum_{j=1}^{J} b_{jn}^2\right) = (\mathbf{X}^{\text{T}}\mathbf{b}_n - \mathbf{y}_n)^{\text{T}}(\mathbf{X}^{\text{T}}\mathbf{b}_n - \mathbf{y}_n) + \lambda\mathbf{b}_n^{\text{T}}\mathbf{b}_n$$

Hint: the derivative of an inner product $(\mathbf{x}^{\text{T}}\mathbf{x})$ with respect to $\mathbf{x}$ can be thought as a vector of derivatives of $(\mathbf{x}^{\text{T}}\mathbf{x})$ with respect to each element of $\mathbf{x}$:

$$d(\mathbf{x}^{\text{T}}\mathbf{x}) = d(x_1{}^2 + x_2{}^2 + \ldots + x_J{}^2) \qquad d(\mathbf{x}^{\text{T}}\mathbf{x})/d\mathbf{x} = \begin{bmatrix} 2x_1 \\ 2x_2 \\ \ldots \\ 2x_J \end{bmatrix} = 2\mathbf{x}$$

References

Araújo, M.C.U., Saldanha, T.C.B., Galvão, R.K.H., Yoneyama, T., Chame, H.C., Visani, V.: The successive projections algorithm for variable selection in spectroscopic multicomponent analysis. Chemom. Intell. Lab. Syst. **57**, 65–73 (2001)

Ben-Gera, I., Norris, K.: Direct spectrophotometric determination of fat and moisture in meat products. J. Food Sci. **33**, 64–67 (1968)

Chung, H., Lee, H., Jun, C.H.: Determination of research octane number using NIR spectral data and ridge regression. Bull. Kor. Chem. Soc. **22**, 37–42 (2001)

Cozzolino, D., Kwiatkowski, M.J., Dambergs, R.G., Cynkar, W.U., Janik, L.J., Skouroumounis, G., Gishen, A.: Analysis of elements in wine using near infrared spectroscopy and partial least squares regression. Talanta. **74**, 711–716 (2008)

Galvão, R.K.H., Araújo, M.C.U., Fragoso, W.D., Silva, E.C., José, G.E., Soares, S.F.C., Paiva, H. M.: A variable elimination method to improve the parsimony of MLR models using the successive projections algorithm. Chemom. Intell. Lab. Syst. **92**, 83–91 (2008)

Golub, G.H., Heath, M., Wahba, G.: Generalized cross-validation as a method for choosing a good ridge parameter. Technometrics. **21**, 215–223 (1979)

Haaland, D.M., Thomas, E.V.: Partial least-squares methods for spectral analysis. 1. Relation to other quantitative calibration methods and the extraction of qualitative information. Anal. Chem. **60**, 1193–1202 (1988)

Hoerl, A.E., Kennard, R.W.: Ridge regression: biased estimation for nonorthogonal problems. Technometrics. **12**, 55–70 (1970)

Kalivas, J.H.: Basis sets for multivariate regression. Anal. Chim. Acta. **428**, 31–40 (2001)

Massart, D.L., Vandeginste, B.G.M., Buydens, L.M.C., De Jong, S., Lewi, P.J., Smeyers-Verbeke, J.: Handbook of Chemometrics and Qualimetrics. Elsevier, Amsterdam (1997)., Chaps. 17 and 36

Norris, K.H., Hart, J.R.: Direct spectrophotometric determination of moisture content of grain and seeds. In: Principles and Methods of Measuring Moisture in Liquids and Solids. Proceedings of the 1963 International Symposium on Humidity and Moisture, vol. 4, pp. 19–25. Reinhold Publishing Co., New York (1965)

Paiva, H.M., Soares, S.F.C., Galvão, R.K.H., Araújo, M.C.U.: A graphical user interface for variable selection employing the successive projections algorithm. Chemom. Intell. Lab. Syst. **118**, 260–266 (2012)

Soares, S.F.C., Gomes, A.A., Araújo, M.C.U., Galvão Filho, A.R., Galvão, R.K.H.: The successive projections algorithm. Trends Anal. Chem. **42**, 84–98 (2013)

Principal Component Analysis

4

Abstract

A brief introduction to principal component analysis is provided, with applications in sample discrimination and in the development of inverse calibration models using full spectral information.

4.1 Why Compressing the Data?

In the end of the previous chapter, the fundamental problem of multivariate calibration was raised: to join in a new model the advantages of CLS (using full-spectral data) and ILS (using inverse calibration). The key appeared to lie in the application of data compressing techniques, in such a way that the original matrix of calibration spectra is reduced in size, preserving its prime information, in the form of sample-to-sample signal changes with constituent concentrations or sample properties. The desired compressed matrix would be a kind of spectral Gulliver in the country of the spectral giants.

By 1960 the compressing technique known as principal component analysis (PCA) was mature. In fact, PCA was first introduced, almost simultaneously in Italy and France, by the mathematicians Eugenio Beltrami and Camille Jordan in 1873 and 1874, respectively. One of the oldest literature references on PCA dates back to the beginning of the twentieth century (Pearson 1901).

PCA generates a matrix $\mathbf{T}$ called the *score* matrix (Fig. 4.1) whose properties will be analyzed in this chapter. The PCA scores efficiently condense the spectral information contained in the real variables within a matrix of appropriate size, to be able to produce a suitable inverse calibration model. This condensation or compression phase of the information contained in $\mathbf{X}$ is essential to understand how the desired model works.

© Springer Nature Switzerland AG 2018

A. C. Olivieri, *Introduction to Multivariate Calibration*,

https://doi.org/10.1007/978-3-319-97097-4_4

Fig. 4.1 Illustrative scheme of the transformation, via PCA, of a large matrix **X** into a smaller score matrix **T**

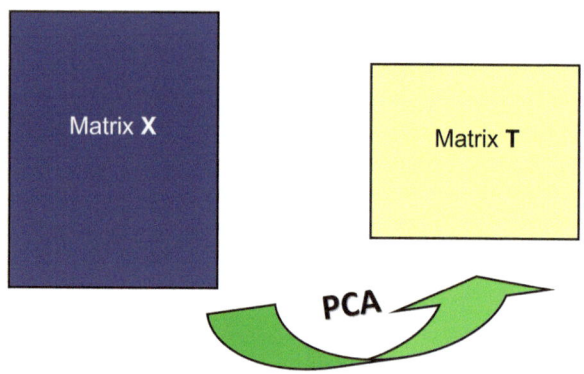

The literature on PCA is abundant (Massart et al. 1997), as it is probably the most employed chemometric technique, not only in chemistry, but in many other scientific fields as well. The purpose of this chapter, therefore, is not to provide a comprehensive insight into PCA, but only to describe in some detail the specific aspects connecting the technique with first-order multivariate calibration.

4.2   Real and Latent Variables

The instrumental signals measured by the analyst are called real or manifest, and differ from the so-called latent variables, originated by a mathematical process from the former ones. They are latent because they cannot be directly visualized in the real experimental signals, unless the latter ones are subjected to mathematical operations which reveal their presence.

The link between real and latent variables is realized through a set of vectors called *loadings*. The loadings provide the connection between the real and the latent worlds when necessary. They are the tools allowing to compress the information, by projecting the latter onto the space represented by them (passage from the real space to the latent one), or to decompress the latent variables to reproduce the original information (passage from the latent space to the real one), as illustrated in Fig. 4.2. This resource will be frequently employed. Each loading corresponds to a given score, so that they come in pairs and are mutually associated. The *loading/score* pair is also called a *principal component*.

4.3   The Principal Components

The first phase of PCA is the computation of loadings. Mathematically, the loadings are the eigenvectors of the square matrix ($\mathbf{X}\,\mathbf{X}^T$) (Watkins 2002). There are several techniques for finding the eigenvectors, such as singular value decomposition (SVD) (Watkins 2002) and NIPALS (*non-linear iterative partial least-squares*) (Wold 1966). The former method estimates all principal components simultaneously,

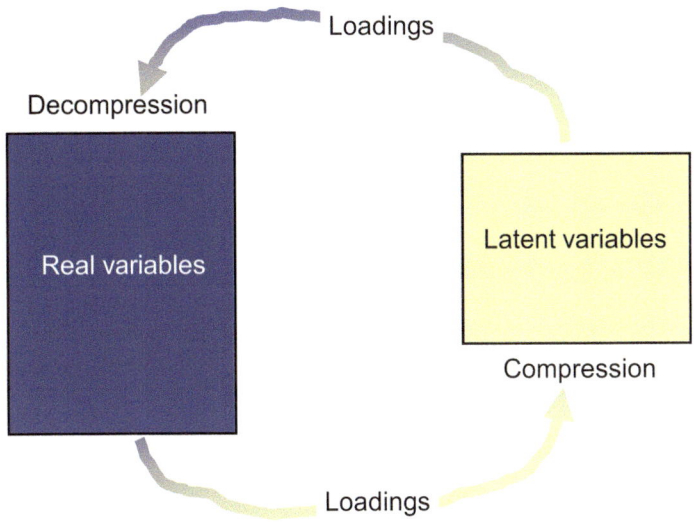

Fig. 4.2 Passage from the real variable space to the latent variable space, by compression or decompression via the spectral loadings

whereas the latter computes them one by one, in the order of the explained proportion of the spectral variations in **X**, until a certain pre-established number of components.

In any case, the loadings are grouped into a matrix **U** (of size $J \times I$). They are called orthonormal, meaning that they are both orthogonal (or perpendicular) and normalized (the length of each loading is unit). These properties can be condensed in the following equation:

$$\mathbf{U}^{\mathsf{T}}\mathbf{U} = \mathbf{I} \qquad\qquad (4.1)$$

where **I** represents an identity matrix of size $I \times I$ (all its diagonal elements are 1 and all off-diagonal ones are 0). The relationship between the original matrix **X**, the loading matrix loadings **U**, and the score matrix **T** is:

$$\mathbf{X} = \mathbf{U}\mathbf{T}^{\mathsf{T}} \qquad\qquad (4.2)$$

Mathematically speaking, one could say that Eq. (4.2) defines a matrix decomposition process. The matrix **X** is decomposed in the product of two matrices, **U** and **T**, on the condition that **U** is formed by orthonormal columns. Figure 4.3 graphically illustrates the process.

To obtain **T** from an experimental data set using Eq. (4.2), we need to carry out these operations: (1) transpose Eq. (4.2) to give $\mathbf{X}^{\mathsf{T}} = \mathbf{T}\,\mathbf{U}^{\mathsf{T}}$, (2) post-multiply by **U**, leading to $\mathbf{X}^{\mathsf{T}}\,\mathbf{U} = \mathbf{T}\,\mathbf{U}^{\mathsf{T}}\,\mathbf{U}$, and (3) note that $(\mathbf{U}^{\mathsf{T}}\,\mathbf{U})$ is the identity matrix (Eq. (4.1)), so that one directly obtains:

$$\mathbf{T} = \mathbf{X}^{\mathsf{T}}\mathbf{U} \qquad\qquad (4.3)$$

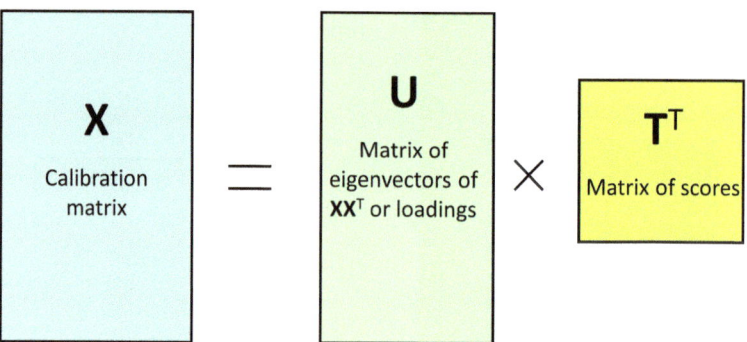

Fig. 4.3 Scheme illustrating the mathematical relationship between the calibration data matrix **X**, the loading matrix **U**, and the score matrix **T**

One could then say that the matrix of scores **T** is the projection of the original data matrix **X** onto the space defined by the loadings contained in **U**. Each column of **T** is a score, and is associated to its corresponding loading or column of **U**. This projection is the fundamental phase of the data compression procedure, because it reduces the size of the original data matrix **X** ($J \times I$) to a score matrix **T** which is in principle smaller ($I \times I$), because usually $J > > I$, i.e., there are much more wavelengths than samples.

One requisite is thus fulfilled, the one related to size. There is another requisite, as important as the former one, which depends on the properties of the columns of **T**. They are also orthogonal, as the columns of **U**. This means that, in a geometrical sense, the columns of **T** are perpendicular to each other, and the correlation among any pair of columns of **T** is zero. Lack of correlation is an important outcome, especially if we recall the discussion around ILS in the previous chapter, where correlations in the signals of the data matrix employed for calibration were a nuisance to any inverse calibration model.

4.4   Highly Significant Loadings and Scores

A judicious analysis of the scores allows one to find that they can be ordered in a consistent manner: according to a decreasing contribution to the spectral variation in **X**. Therefore, if one not only wishes to compress **X** to its reduced version **T**, but also to select inside **T** the relevant information, separating it from the irrelevant one, it is possible to further reduce the size of **T**. This even more reduced version of **T** is called *truncated*, because it is not obtained by projection onto the loading space, but directly by pruning the irrelevant **T** columns.

Suppose we have measured the spectra of 100 experimental samples at 1000 wavelengths. The matrix **X** will be of size 1000×100, containing 100,000 numbers. On the other hand, if in the **T** matrix (of size 100×100) we discover that only two columns are highly significant, we could then represent most of the variation in **X** by

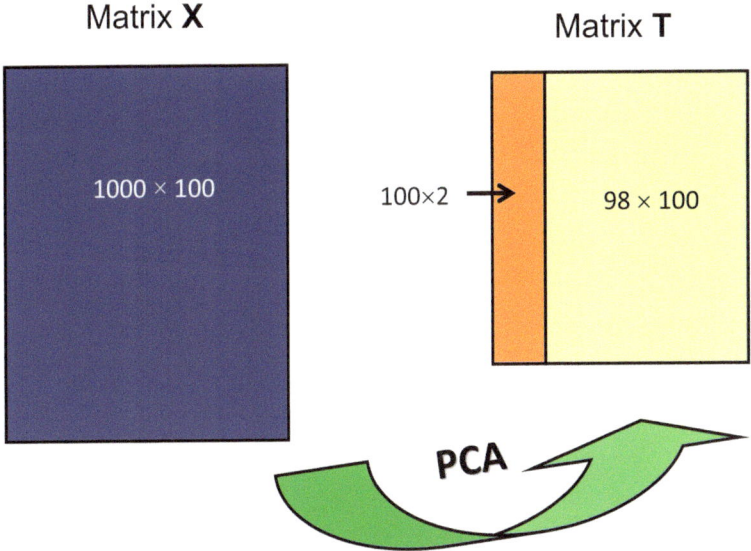

Fig. 4.4 Illustrative scheme of the transformation, via PCA, of a matrix **X** of size 1000×100 into a smaller score matrix **T**, of size 100×100. The first two columns of **T** are highly significant, while the remaining 98 mainly represent instrumental noise

a truncated matrix of size 100×2, with only 200 numbers: those in the two highly significant columns of **T**. The degree of compression is amazing: the matrix size is reduced by 99.8%, leaving a matrix **T** with only 0.2% of the original size. This is the sense of compression: to collect the valuable information into a few numbers. In this example there will be two highly significant principal components, with two scores and two associated loadings. In the consistently ordered versions of **T** and **U**, the highly significant scores and loadings will be the first two columns of **T** and the first two columns of **U**. The situation is schematically represented in Fig. 4.4.

4.5 Poorly Significant Loadings and Scores

If the previously described matrix **T** has 100 columns and only two are found to be highly significant, what do the 98 remaining columns represent? Answer: they are columns of scores which are primarily related to spectral noise. If we could spy the specific numbers contained in the loading matrix **U**, we would see that each column has J elements, as many as wavelengths. We could then speak about the spectra of the loadings, and plot them as a function of wavelength. A highly significant spectral loading (one of the first two columns of **U** in the above example) would show details which are typically spectral, although with positive and negative features, meaning that they do not represent true constituent spectra. They would instead be linear

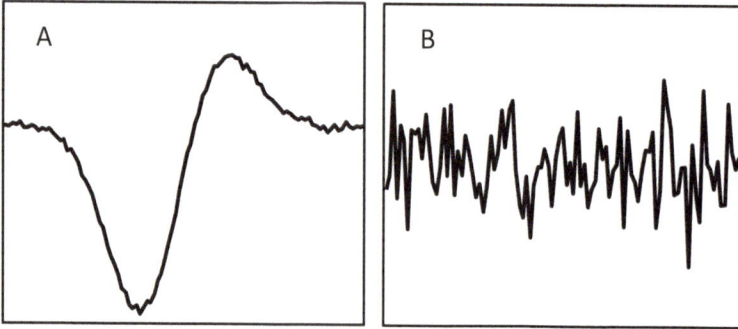

Fig. 4.5 (**a**) Plot of a typical highly significant spectral loading. (**b**) Plot of a typical poorly significant spectral loading

Fig. 4.6 The specific content of the matrix **U**. The first two columns have spectral features, whereas the subsequent ones are random noise

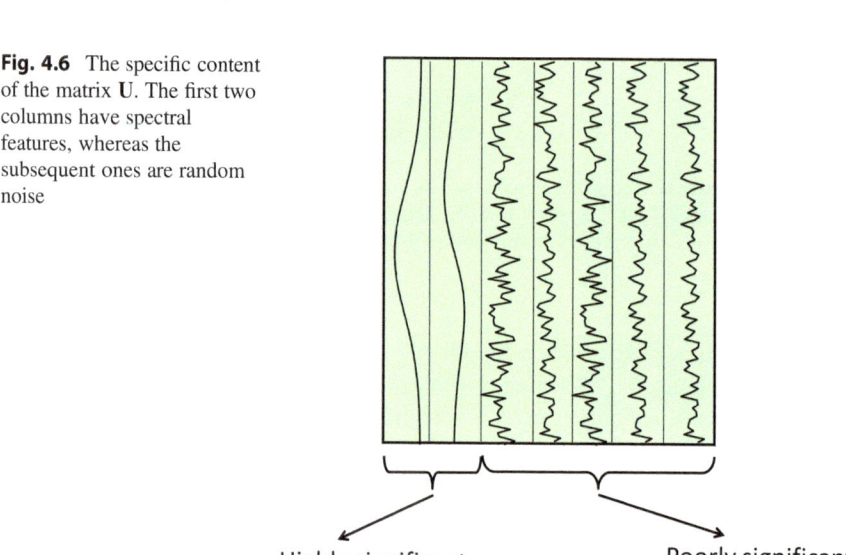

combinations of those pure spectra, with the meaning of catching, in a purely mathematical/statistical sense, the spectral variability shown by **X**. On the other hand, the poorly significant loadings would show spectral random noise, precisely because they model the instrumental noise present in **X**. This is illustrated in Fig. 4.5.

In the ideal case, the loadings will be highly significant up to a certain point, beyond which the noise-related loadings will start to appear (Fig. 4.6). In real life, as one could easily anticipate, the situation will not be a black-and-white one, and there would be some intermediate loadings. Visual inspection of the latter may not be able to classify them as either spectral or random. Here is where mathematical methods may help us in deciding, on a statistical basis, how many loadings are really significant in a given matrix **U**. We will return on this very important matter in the future.

4.6   Application to Sample Discrimination

Principal component analysis has a relevant application in practice, in the discrimination of samples by means of spectroscopy or other instrumental techniques. Suppose the matrix **T** of size 100×100 was truncated to a size 100×2, retaining only the first two columns, because they are the highly significant ones in comparison with the remaining 98 columns. If such is the case, each column of the truncated matrix contains two numbers for each of the 100 samples: the first and second scores, as illustrated in Fig. 4.7.

We could plot the location of each sample, taking as x coordinate the value of the first score and as y coordinate the value of the second score. This may result in a map such as the one of Fig. 4.8 where samples belonging to a certain class (A) are separated, in the score space, from samples belonging to another class (B). Notice that the separation is obtained without previously knowing that there are two sample classes. This type of analysis is known as *unsupervised*, because samples are

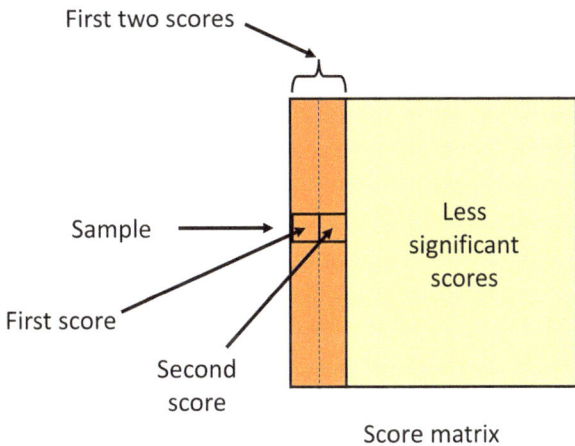

Fig. 4.7 Scheme showing how the elements of the first two columns of the score matrix **T** can be associated to a specific sample of a set

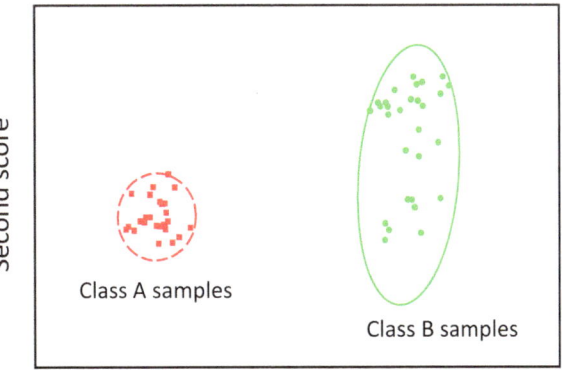

Fig. 4.8 Plot of the second score vs. the first score for a set of samples belonging to two different classes

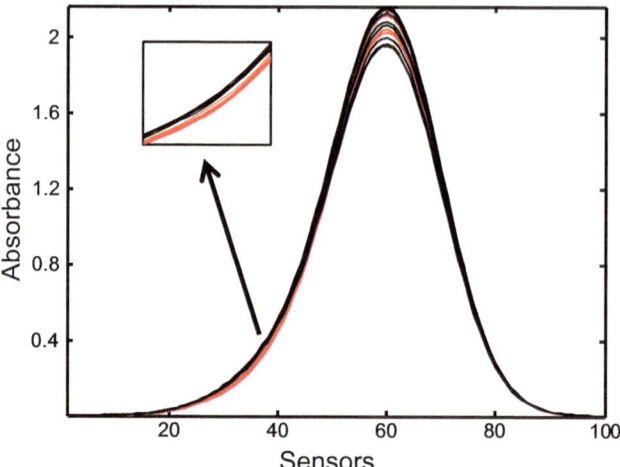

Fig. 4.9 Spectra of 20 samples at 100 wavelengths (red lines, samples of class A; black lines, samples of class B). The insert shows a zoom of the region between sensors 30 and 40

separated in an automatic manner, without previous knowledge on their origin. In the *supervised* analysis, on the other hand, one previously know the sample classes, and this information is employed, together with the spectra or other multivariate signals, to train a suitable discrimination model.

The best preacher is brother example (an old Spanish proverb). Figure 4.9 shows the spectra for 20 different samples measured at 100 wavelengths. The samples are suspected to belong to two different classes: 10 to class A and 10 to class B, and the experiment is set to investigate whether: (1) they can be discriminated on the basis of their spectra and (2) if there is a particular spectral region which is responsible for the discrimination, meaning that in that region the constituents responsible for the discrimination absorb.

The result of applying PCA to the spectra is a set of loadings and associated scores. The first option in this type of studies is always to place the samples in a two-dimensional plot, in which a given point corresponds to a sample, and is defined by the first and second score. The plot is shown in Fig. 4.10, where a reasonably good separation is apparent in two sub-sets of 10 samples each. It is important to notice that the second score is the parameter which is really useful for discrimination of the samples in two types (Fig. 4.10), because the samples with large values of the second score belong to one class, whereas those with low values of the second score belong to the other class, almost independently on the value of the first score.[1]

---

[1]The separation of a sample set into classes by PCA is called *discrimination*, as described here. The term *classification* is reserved for the development of a rule for assigning future samples to any of the separated classes. For example, in Fig. 4.10 the rule might be: samples with positive second score belong to one class and samples with negative second score to the other class.

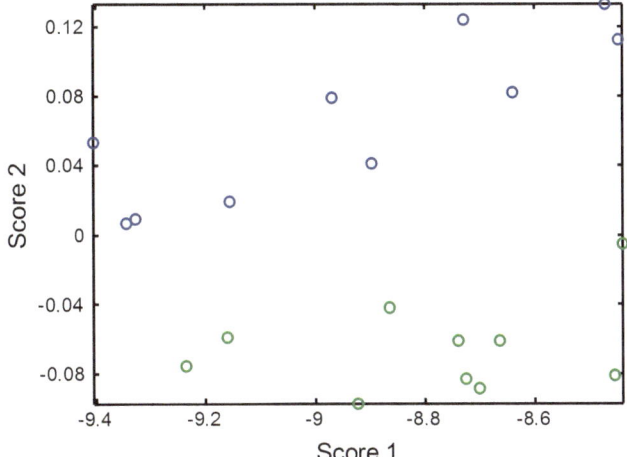

Fig. 4.10 Plot of the second score vs. the first score, after PCA of the spectra of Fig. 4.9. Blue circles, samples of class A, green circles, samples of class B

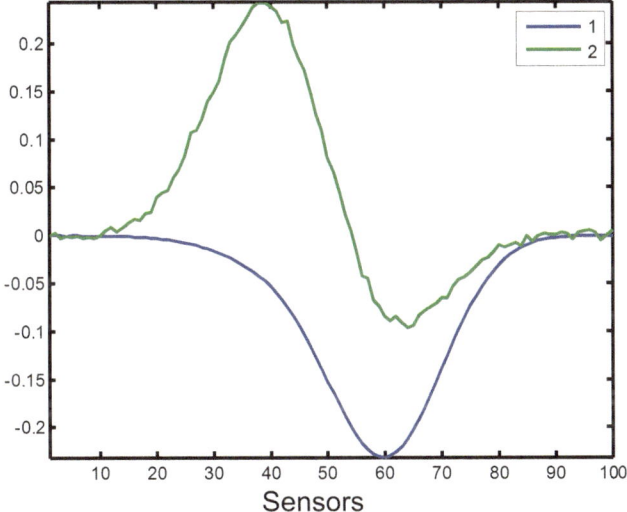

Fig. 4.11 First two loadings after PCA of the spectra of Fig. 4.9

Another important phase of PCA is the study of the loadings. Figure 4.11 shows the first two loadings, which have apparent spectral aspect, although they are abstract linear combinations of real spectra. What is the connection between Figs. 4.10 and 4.11? If the second score is fundamentally responsible for the successful discrimination, then the second loading should show high intensities in the spectral regions where the responsible constituents absorb. It is apparent that the spectral region in the range of sensors 30–40 is the responsible one. We would then search the

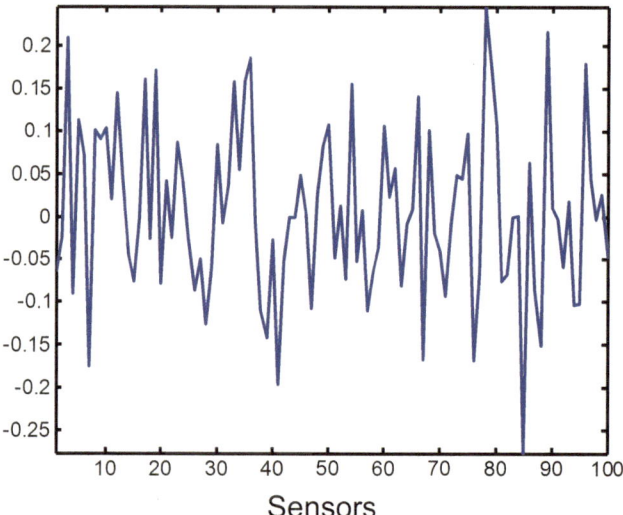

Fig. 4.12   Third loading after PCA of the spectra of Fig. 4.9

chemical structures likely to absorb in this region, verifying if they are really responsible for the discrimination. Going back to Fig. 4.9 and inspecting it in detail, we discover that in the region 30–40 sensors there is a small spectral shoulder where samples differentiate, although minimally, from one type to another (see the insert zooming that particular sensor region). PCA has made completely apparent this fact that could go unnoticed to the naked eye. Some authors rightly affirm that PCA is a tool allowing one to find hidden patterns in the data. That small shoulder is the hidden feature revealed by PCA in the present case.

If the second principal component is the discriminatory one, we could ask ourselves what is the role of the first one. Comparison of Figs. 4.9 and 4.11 suggests that the first loading is related to the mean spectrum of the set of original spectra (multiplied by -1), and does not provide relevant information regarding the discrimination. For this reason, it is usual to employ a mathematical transformation of the raw data previous to PCA, consisting in centering the spectra with respect to the mean (subtracting from each spectrum the mean spectrum). This will decrease the contribution of the mean spectrum, and will highlight the role of the spectral differences leading to discrimination. We will return on this activity in the future.

We finally study the subsequent loadings, from the third to the last one. Figure 4.12 plots the third loading, which is apparently random noise. Likewise, all the subsequent loadings are composed of random noise. This means that the remaining scores, associated to these loadings, do not contain information useful for discrimination. We can tie this behavior to the numerical values of the portion of the data which is explained by each principal component. The latter is usually called the *explained variance*, and is estimated by writing Eq. (4.2) as follows:

$$\mathbf{X} = \mathbf{U}\mathbf{T}^{\mathrm{T}} = [\mathbf{u}_1 \ \mathbf{u}_2 \ \ldots \ \mathbf{u}_I] \begin{bmatrix} \mathbf{t}_1^{\mathrm{T}} \\ \mathbf{t}_2^{\mathrm{T}} \\ \ldots \\ \mathbf{t}_I^{\mathrm{T}} \end{bmatrix} = \mathbf{u}_1 \mathbf{t}_1^{\mathrm{T}} + \mathbf{u}_2 \mathbf{t}_2^{\mathrm{T}} + \ldots + \mathbf{u}_I \mathbf{t}_I^{\mathrm{T}} \quad (4.4)$$

where the columns $\mathbf{u}_1$, etc. are the loadings and the rows $\mathbf{t}_1^{\mathrm{T}}$, etc. the scores. The above equation allows one to consider that each term of the form $(\mathbf{u}_i \mathbf{t}_i^{\mathrm{T}})$ contributes to the reconstruction of $\mathbf{X}$ with a certain proportion of the data. The specific contribution of the successive terms is usually measured as the sum of their squared elements, relative to the sum of all the squared elements of $\mathbf{X}$. This defines the explained variance by each principal component as:

$$\text{Explained variance}\,(\%) = 100 \frac{\sum\limits_{j=1}^{J} \sum\limits_{i=1}^{I} \left(u_j t_i\right)^2}{\sum\limits_{j=1}^{J} \sum\limits_{i=1}^{I} x_{ji}^2} \quad (4.5)$$

where x_{ji}, u_j, and t_i are generic elements of $\mathbf{X}$, $\mathbf{u}$, and $\mathbf{t}$, respectively.

In the present case, the explained variances are: 98.4% by the first principal component, 1.0% by the second and 0.6% by the sum of all the remaining ones, or ca. 0.03% each. This justifies why only two principal components are able to explain the main fraction of the spectral variability. Notice that the discrimination was made possible thanks to that 1% explained by the second component.

In a real case, it may not be possible to truncate the matrix $\mathbf{T}$ to two columns only, because more columns are associated to relevant phenomena, so that sample discrimination may require more components per sample. If the first three principal components are highly significant, then a three-dimensional plot could be useful to separate the samples.

4.7 A PCA Algorithm

Box 4.1 gives a short MATLAB code for implementing PCA, obtaining the loadings and scores, as well as the explained variance by each principal component.

Box 4.1
This PCA algorithm invokes a sub-routine of the MATLAB environment ('princomp'), and provides the loadings and scores from an input variable 'X', the matrix of spectra of size $J \times I$ (J = number of wavelengths or sensors, I = number of calibration samples)

```
[U,T,L]=princomp(X','econ') ;
EV=100*L/sum(L);
```

(continued)

Box 4.1 (continued)

The output consists of 'U', the loadings, 'T', the scores, and 'EV', the explained variance by each component. If one wants to plot the first two loadings and the second vs. the first score, the commands are:

        plot(U(:,1:2))

        plot(T(:,1),T(:,2),'o')

Notice that the MATLAB command 'princomp' centers the data before applying PCA.

4.8   A Real Case

A literature work describes the discrimination of tea samples according to its variety (He et al. 2007). A total of 240 samples of eight typical kinds of tea were purchased at a local super-market: Zisun, Xihu Longjing, Zhejiang Longjing, Yangyangouqin, Xushuiyunlv, Maofeng, Lushanyunwu, and Wanhai. Figure 4.13 shows the NIR spectra of the tea samples (He et al. 2007). The data were not submitted to PCA directly, but were previously transformed using a mathematical procedure to remove the dispersion effects of the NIR radiation when studying solid samples. These pre-processing methods will be discussed in detail in a future chapter.

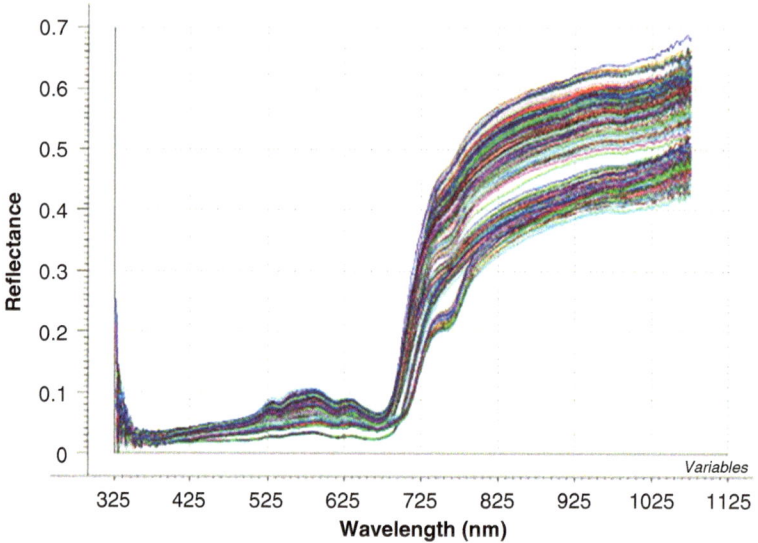

Fig. 4.13 NIR spectra of samples of eight different varieties of tea. Reproduced with permission from Elsevier (refer Footnote 1)

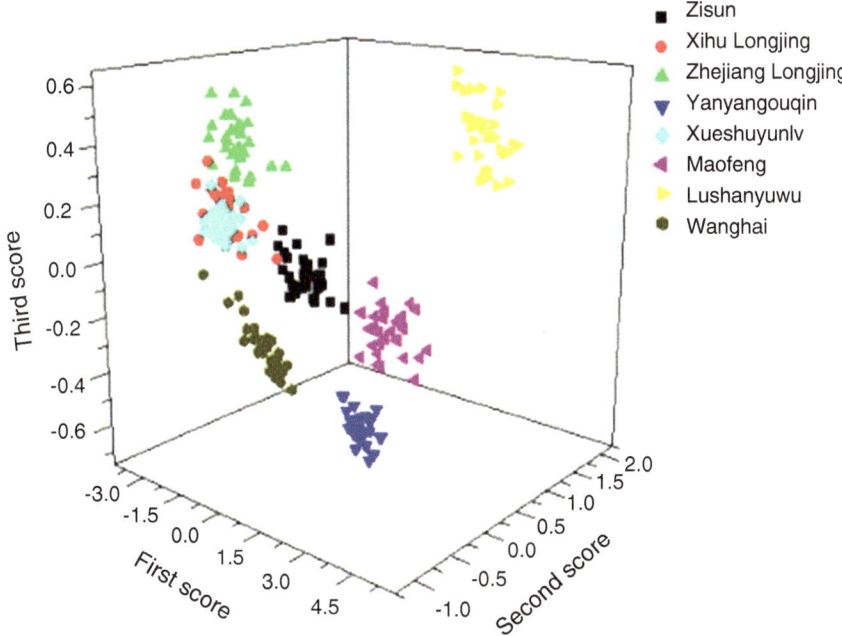

Fig. 4.14 Three-dimensional plot of the first, second, and third scores of the tea samples, as indicated. Reproduced with permission from Elsevier (refer Footnote 1)

After applying PCA, the authors found that three principal components were needed to explain the spectral variance, and were all involved in the discrimination process. Figure 4.14 shows the corresponding three-dimensional score plot, where the discrimination success is apparent.

> In the authors' words: ... *The contribution of this work is to present a rapid and non-destructive approach for discriminating of different varieties of tea. At present, there are only qualitative analysis in most of the discrimination of varieties, ... In this research, we made quantitative analysis for the varieties of tea, ... a relation was established between reflectance spectra and varieties of tea* ... (He et al. 2007).

4.9 Application to Multivariate Calibration

The main application of PCA in the framework of multivariate calibration is to provide a truncated score matrix, of adequate size and properties, to be coupled to an inverse calibration model. This would create a model capable of employing full-spectral data to quantitate analyte concentrations or global sample properties.

The size of the truncated score matrix will be $I \times A$, where I is the number of calibration samples, and A is the number of columns of $\mathbf{T}$ explaining the largest

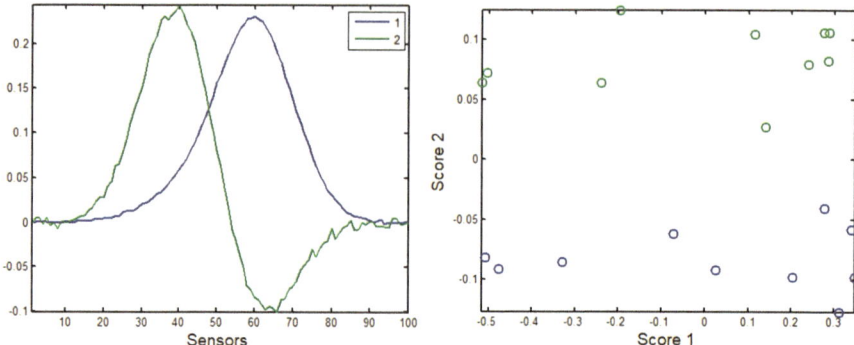

Fig. 4.15 Left, first two loadings of the spectra discussed in Sect. 4.6 after centering the data matrix with respect to the mean spectrum. Right, plot of second vs. first scores after mean centering the data

proportion of the variance, which are associated to the highly significant loadings. It is not possible to exaggerate how crucial is the estimation of the optimum value of A in multivariate calibration. There are various alternative procedures, visual and statistical, for this important purpose. We will devote an entire chapter to the subject.

On the other hand, as regards the properties of the score matrix, its columns are orthogonal to each other. The most important consequence of this property is that this matrix is free from collinearity or overlapping among its columns. This will be highly valuable in the next chapter.

4.10   Exercises

1. Figure 4.15 shows the loadings (A) and a plot of second vs. first scores (B) for the simulated problem discussed in Sect. 4.6, after centering the spectra with respect to the mean spectrum. The explained variances by the first and second principal components are 69.0% and 18.6%, respectively.
   (a) What conclusions can be drawn from the explained variances with respect to the use of raw (uncentered) data?
   (b) Compare the loadings in Fig. 4.15 A with those for uncentered data
   (c) Compare the score–score plot in Fig. 4.15 B with that for uncentered data

References

He, Y., Li, X., Deng, X.: Discrimination of varieties of tea using near infrared spectroscopy by principal component analysis and BP model. J. Food Eng. **79**, 1238–1242 (2007)

Massart, D.L., Vandeginste, B.G.M., Buydens, L.M.C., De Jong, S., Lewi, P.J., Smeyers-Verbeke, J.: Handbook of Chemometrics and Qualimetrics. Elsevier, Amsterdam (1997). Chapter 31

Pearson, K.: On lines and planes of closest fit to systems of points in space. Phil. Mag. **2**, 559–572 (1901)

Watkins, D.S.: The singular value decomposition (SVD). In: Fundamentals of Matrix Computations, 2nd edn. Wiley, Hoboken, NJ (2002)

Wold, H.: Estimation of principal components and related models by iterative least squares. In: Krishnaiah, P.R. (ed.) Multivariate Analysis, pp. 391–420. Academic Press, New York (1966)

Principal Component Regression

5

Abstract

A modern multivariate model incorporating all required characteristics is discussed, based on the combination of principal component analysis and inverse least-squares regression.

5.1 PCA and ILS Combined: Another Brilliant Idea

When considering the CLS and ILS models, the important question was: why not exploiting the advantages of both models? At the end of Chap. 3 this question was raised, proposing a working philosophy to overcome the problems of CLS and ILS. The proposal consisted in estimating a new matrix, from the original one of instrumental signals for calibration $\mathbf{X}$, preserving the information regarding the chemical constituents and spectra that were latent in $\mathbf{X}$. On the other hand, the size of the new matrix should be considerably smaller than that of $\mathbf{X}$, compatible with the requirements of the ILS model. By the time ILS was first proposed (the decade of 1960), a technique to produce the desired matrix was long known. The latter is the truncated score matrix furnished by principal component analysis (PCA), as studied in the previous chapter.

The combination of PCA and ILS gives rise to the first-order multivariate model known as principal component regression (PCR), and represents one of the simplest attempts to integrate the main advantages of CLS and ILS. The PCR model employs an inverse calibration, but does not correlate the analyte concentrations or sample properties with the instrumental responses, but with the truncated score matrix discussed in the previous chapter. As before, we highly recommend reading the work of Haaland and Thomas on the subject (Thomas and Haaland 1988).

© Springer Nature Switzerland AG 2018

A. C. Olivieri, *Introduction to Multivariate Calibration*,

https://doi.org/10.1007/978-3-319-97097-4_5

5.2   Matrix Compression and Decompression

The main idea of PCR is to replace the original calibration data matrix $\mathbf{X}$ by a compressed version. The replacement is the truncated version of the score matrix $\mathbf{T}$, which only retains the first A columns. We will call the truncated score matrix $\mathbf{T}_A$. This matrix, of size $I \times A$, is composed of the A mutually orthogonal columns of $\mathbf{T}$ explaining a significant portion of the spectral variance, associated, as we have seen before, to the A highly significant loadings.

We are thus assuming that these A latent variables explain the variance in $\mathbf{X}$, and that it is not necessary to employ the complete matrix $\mathbf{U}$ in the projection of $\mathbf{X}$ to find $\mathbf{T}$ (see Chap. 4). One can remove the columns of $\mathbf{U}$ from the $(A + 1)$ to the I, leaving a matrix $\mathbf{U}_A$ (of size $J \times A$) only containing the first A loadings. The remaining loadings are discarded because they are considered to model the instrumental noise. In this way, the truncated score matrix $\mathbf{T}_A$ can be expressed as:

$$\mathbf{T}_A = \mathbf{X}^\mathrm{T}\mathbf{U}_A \tag{5.1}$$

This matrix $\mathbf{T}_A$, in spite of having a size considerably smaller than the original spectral matrix, plays a similar role in calibration, because the relevant information present in $\mathbf{X}$ has been compressed and selected in an efficient way.

On the other hand, given a pair of truncated matrices $\mathbf{T}_A$ and $\mathbf{U}_A$, we can reconstruct an approximation to $\mathbf{X}$ which we call $\mathbf{X}_A$ by means of:

$$\mathbf{X}_A = \mathbf{U}_A\mathbf{T}_A{}^\mathrm{T} \tag{5.2}$$

This latter matrix $\mathbf{X}_A$ is a different version of $\mathbf{X}$: it is an approximation to $\mathbf{X}$ reconstructed only with the really useful information, discarding the instrumental noise. The process can be illustrated as a series of images of a Brazilian beach in Fig. 5.1. The original picture is a matrix of pixel intensities, which can be decomposed using PCA. The picture can then be reconstructed using Eq. (5.2). The image for $A = 1$ was reconstructed using only the first principal component, which is the one retaining the largest portion of the total variance. As more and more principal components are employed in the reconstruction of $\mathbf{X}_A$, the image becomes progressively neat, but the relevant information is retained by the compressed matrices, even when using a reduced number of latent variables. In the specific case of Fig. 5.1, the original picture has 5,308,416 data values, whereas the one reconstructed with 50 latent variables required 268,850 intensities, implying a compression of ca. 96%.

Figures 5.2 and 5.3 graphically show the operations of: (1) matrix decomposition of $\mathbf{X}$ into loadings and scores, and (2) reconstruction of the matrix $\mathbf{X}_A$ from the truncated score and loading matrices.

Original image

Reconstructed images

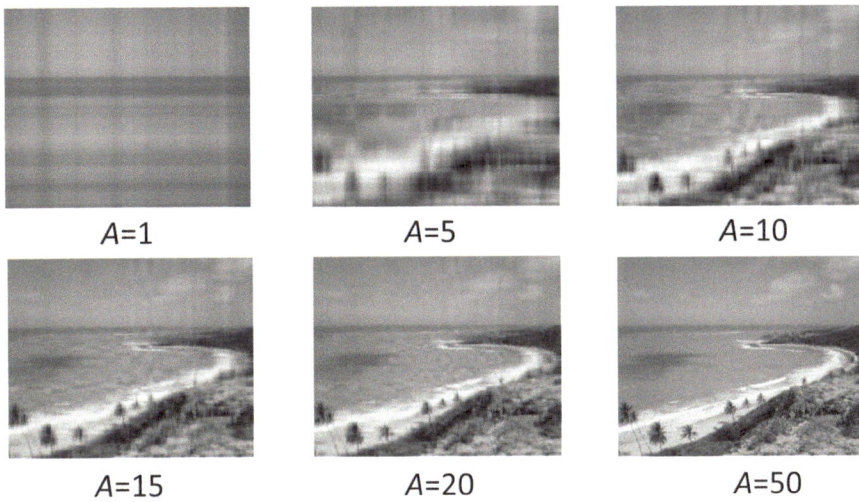

A=1 A=5 A=10

A=15 A=20 A=50

Fig. 5.1 Reconstruction of a digital image using a limited number of principal components (indicated by the value of A). The original image has been taken by the author in a Brazilian beach

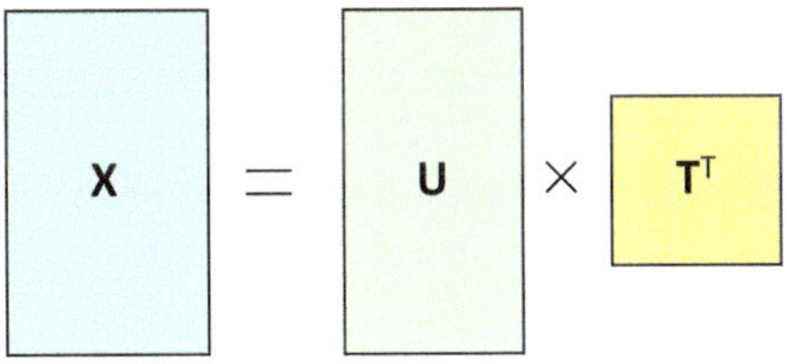

Fig. 5.2 Graphical representation of the decomposition of the original data matrix **X** in the product of the loading and score matrices

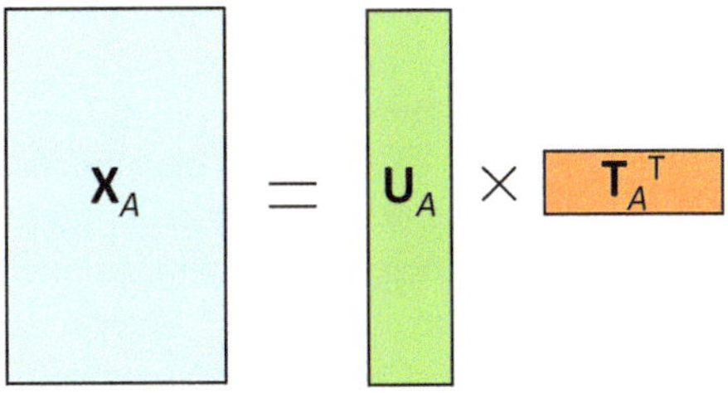

$$\mathbf{X} = \mathbf{U} \times \mathbf{T}^{\mathsf{T}}$$

Fig. 5.3 Graphical representation of the reconstruction of the matrix $\mathbf{X}_A$ with the product of the truncated loading $\mathbf{U}_A$ and score $\mathbf{T}_A$ matrices, using a reduced number of principal components (A)

$$\mathbf{X}_A = \mathbf{U}_A \times \mathbf{T}_A^{\mathsf{T}}$$

5.3   Calibration Phase

At this point we join the advantages of CLS and ILS, which was the prime objective of multivariate calibration. An inverse calibration model is built, in which the analyte concentrations in the training samples contained in $\mathbf{y}_n$ are correlated with the scores contained in the truncated score matrix $\mathbf{T}_A$, instead of the original data matrix $\mathbf{X}$:

$$\mathbf{y}_n = \mathbf{T}_A \mathbf{v}_n + \mathbf{e} \qquad (5.3)$$

where $\mathbf{v}_n$ (of size $A \times 1$) is a vector of regression coefficients defined in the space of the latent variables and $\mathbf{e}$ collects the error models. The vector $\mathbf{v}_n$ can be found from Eq. (5.3) pre-multiplying both sides by $\mathbf{T}_A{}^T$:

$$\mathbf{T}_A{}^T \mathbf{y}_n = \mathbf{T}_A{}^T \mathbf{T}_A \mathbf{v}_n + \mathbf{e} \qquad (5.4)$$

We now multiply both sides by the inverse of $(\mathbf{T}_A{}^T \mathbf{T}_A)$ and get $\mathbf{v}_n$:

$$\mathbf{v}_n = (\mathbf{T}_A{}^T \mathbf{T}_A)^{-1} \mathbf{T}_A \mathbf{y}_n \qquad (5.5)$$

By analogy with the criterion employed before, we call the matrix $[(\mathbf{T}_A{}^T \mathbf{T}_A)^{-1} \mathbf{T}_A]$ as the generalized inverse of $\mathbf{T}_A$ and represent it by $\mathbf{T}_A{}^+$, so that Eq. (5.5) adopts the following final form:

$$\mathbf{v}_n = \mathbf{T}_A{}^+ \mathbf{y}_n \qquad (5.6)$$

This step completes the calibration. Obtaining the regression coefficients $\mathbf{v}_n$ in PCR is completely analogous to the process of finding the regression coefficients $\mathbf{b}_n$ in CLS. The only difference is that $\mathbf{v}_n$ is defined in a latent space, a highly reduced and abstract version of the real space. Just to set an example, if a CLS model is built with spectra measured at 1000 wavelengths, the vector of regression coefficients $\mathbf{b}_n$ will be a column vector with 1000 elements, whereas if only two principal components retain the variance, and the score matrix is truncated to its first two columns, the vector of regression coefficients in PCR $\mathbf{v}_n$ will only have two elements. This implies a reduction of 99.8% in the number of coefficients, leaving only 0.2% of the latent counterparts.

5.4   Mathematical Requirements

In the ILS calibration phase we found, as the main drawback, the issue of inverting the square matrix $(\mathbf{X} \mathbf{X}^T)$. The mathematical requirements for the latter inversion to be possible were: (1) $I > J$, that is, more samples than wavelengths and (2) low degree of correlation among columns of $\mathbf{X}$.

As regards the number of independent equations and unknowns, Eq. (5.3) represents a system of multiple equations (the number of equations is I, equal to the number of samples or elements of $\mathbf{y}_n$) with multiple unknowns (the number is A, equal to the elements of $\mathbf{v}_n$). This implies that the number of calibration samples should be larger than the number of columns of the truncated matrix $\mathbf{T}_A$. This is indeed fulfilled, because the maximum possible value of A is I.

Notice that A represents the number of sources of spectral variance present in the system. A rule of thumb of PCR calibration is that as the analytical systems become more complex, they will show a larger number of sources of variance (mostly chemical constituents, but also baseline drifts, dispersive effects of the radiation, etc.) and will require a correspondingly larger number of calibration samples. Calibration developers for analytes or properties in industrial or natural samples are used to collect large sets of calibration samples, on the order of hundreds or thousands. We may naturally expect that, in these systems, $A \ll I$.

In the case of PCR, the inversion of $(\mathbf{T}_A{}^\mathrm{T} \mathbf{T}_A)$ in Eq. (5.5) is trivial, because this matrix is diagonal (all off-diagonal elements are zero), due to the fact that the columns of $\mathbf{T}_A$ are orthogonal. The inverse of a diagonal matrix is simply given by a diagonal matrix whose elements are the inverse of the original ones. This greatly simplifies the collinearity issues which were so problematic in ILS.

5.5   Prediction and Validation Phases

In the prediction phase, the spectrum of an unknown sample is registered, and the instrumental signals are collected in the column vector $\mathbf{x}$ (of size $J \times 1$). Before applying the prediction model, it is necessary to project the latter vector onto the space of the A columns of the truncated loading matrix $\mathbf{U}_A$, because we cannot use the original data to estimate concentrations mixing a real spectral vector $\mathbf{x}$ with the compressed regression coefficients contained in $\mathbf{v}_n$.

In an analogous fashion to Eq. (5.1), a score vector $\mathbf{t}_A$ (of size $A \times 1$) is obtained for the unknown sample:

$$\mathbf{t}_A = \mathbf{U}_A{}^\mathrm{T}\mathbf{x} \tag{5.7}$$

This vector $\mathbf{t}_A$ contains the sample scores. They will play the role of the real spectral data in the prediction PCR phase, where the inverse model is applied (concentration proportional to signal):

$$y_n = \mathbf{v}_n{}^\mathrm{T}\mathbf{t}_A \tag{5.8}$$

where $\mathbf{v}_n$ replaces $\mathbf{b}_n$ and $\mathbf{t}_A$ replaces $\mathbf{x}$. From the last equation, the analyte concentration or sample property y_n can be estimated.

Validation, in turn, proceeds as for the previous CLS and ILS models.

5.6   The Vector of Regression Coefficients

Equation (5.8) is a highly compressed version of the prediction phase. We would like to go back to the real space, and get a vector of regression coefficients $\mathbf{b}_n$ defined in the original wavelength space. In this way, we would get *the equation*, the product that some calibration developers sell to the customers. *The equation* is the set of numbers, one for each wavelength, allowing one to estimate the analyte concentration in the sample, multiplying the original spectrum $\mathbf{x}$.

As commented above, the link between the latent and the real space is the loading matrix, in this case the truncated loading matrix $\mathbf{U}_A$. Indeed, replacing $\mathbf{t}_A$ from Eq. (5.7) in Eq. (5.8):

$$y_n = \mathbf{v}_n^{\mathrm{T}} \mathbf{U}_A^{\mathrm{T}} \mathbf{x} \qquad (5.9)$$

We can verify that the vector of regression coefficients in the real space $\mathbf{b}_n$ (transposed) is the product $\mathbf{v}_n^{\mathrm{T}} \, \mathbf{U}_A^{\mathrm{T}}$, so that:

$$\mathbf{b}_n = \left(\mathbf{v}_n^{\mathrm{T}} \mathbf{U}_A^{\mathrm{T}} \right)^{\mathrm{T}} = \mathbf{U}_A \mathbf{v}_n \qquad (5.10)$$

We can appreciate here the usefulness of the truncated loading matrix, in allowing us to decompress the latent regression vector to the real space, and to find a vector of regression coefficients analogous to the one for CLS, but obtained in the framework of an inverse compressed model.

Recall the discussion on the CLS vector of regression coefficients in Chap. 2, Sect. 2.6. We noticed that the spectrum of $\mathbf{b}_n$ showed features indicating where the analyte of interest responded. The system was a simulated one with well-resolved pure constituent spectra, and the model was a classical, direct one, resorting to Lambert–Beer's law. Does the PCR regression vector $\mathbf{b}_n$ of Eq. (5.10) represent something physical too? To answer this question let us describe an experimental example. The NIR spectra of 50 samples of whole corn seeds of known moisture level were measured, the number of latent variables A for truncating the loading matrix was estimated, and a PCR model was built to predict the level of moisture from the spectra of future samples. Figure 5.4 shows the spectra, and Fig. 5.5 the vector of PCR regression coefficients $\mathbf{b}_n$. As can be seen, the latter presents an intense band at ca. 1900 nm. Does this band represent a real phenomenon, in spite of the abstract way in which the regression coefficients were found? Think about the process: principal component analysis through the eigenvectors of the spectral data matrix, truncation of the score matrix, inverse calibration for the moisture values, and decompression of the latent vector of regression coefficients back to the real space. Honestly, it will be rather miraculous if Fig. 5.5 carries any physical sense.

However, *it does*. The PCR calibration of the moisture level should in principle be related to the NIR absorption properties of the water molecule. Water shows various absorption bands in its NIR spectrum; the most intense one is located at ca. 1900 nm, and corresponds to a frequency of 5260 cm^{-1}, which is a combination band consisting of the sum of two fundamental vibrational frequencies: stretching

Fig. 5.4 NIR spectra of corn seeds, employed to build a PCR model for the rapid determination of the moisture level and other important parameters. They were measured in the wavelength range 1100–2498 nm at 2 nm intervals (700 channels), and are available at http://www.eigenvector.com/data/Corn

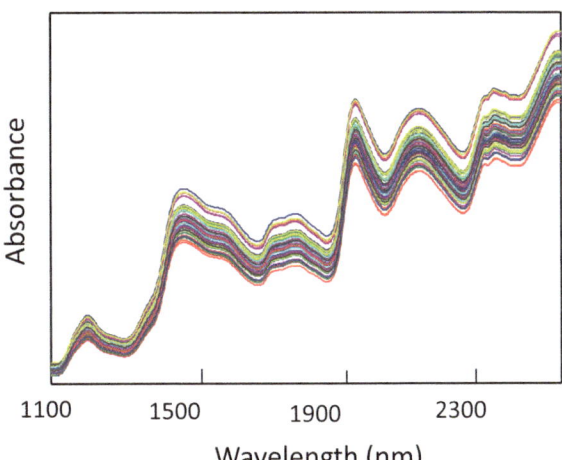

Fig. 5.5 PCR regression coefficients for the calibration of moisture in corn seeds. The wavelengths for the major peaks are indicated

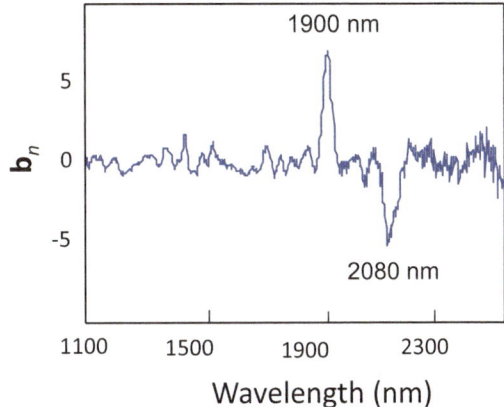

(ca. 3700 cm^{-1}) and bending (ca. 1600 cm^{-1}). Figure 5.5 shows that the most intense positive band in the vector of PCR regression coefficients is located at ca. 1900 nm. This cannot be a chance result.

The lesson from this section is simple but extremely important: the vector of regression coefficients *knows* where the analyte responds in the spectra of the calibration samples, or where the chemical constituents responsible for an organoleptic property of a sample respond. Even after a complex set of operations such as sample collection, spectral measurement, and mathematical processing, the vector of regression coefficients will be there, eager to provide a physicochemical interpretation of the phenomenon under study. This implies that calibration models, no matter how complex in their formulation, have always a connection with real phenomena, about which the analyst may have some experience. A PCR model providing a vector $\mathbf{b}_n$ which is only random noise cannot be a good predictive model.

In sum, spectral bands which are known to be intense in the pure analyte spectrum (provided this information is available) may in principle be apparent in the spectrum of $\mathbf{b}_n$. However, this may not always be the case, for a simple reason: if some analyte bands are severely overlapped with those for other constituents acting as interferents, there will be no net signal left for prediction, and $\mathbf{b}_n$ will be small, possibly noise in the position of those specific bands. All this means that the physical interpretation of $\mathbf{b}_n$ may not be completely transparent.

5.7   Karl Norris and the Regression Coefficients

The work that started the modern era of multivariate calibration is due to Norris, as previously discussed (Norris and Hart 1965). He was able to show that the moisture level of various samples (wheat, soybean, wheat flour, and wheat bran) could be reasonably predicted by a very simple model. Specifically, he found that only measurements at two NIR wavelengths were required, and that the moisture (y) was directly proportional to the difference in absorbance at 1940 and 2080 nm, so that $y = k\,(x_{1940} - x_{2080})$. He did not use PCR, but only ILS at these two wavelengths.

Figure 5.5, on the other hand, was obtained after a considerably more elaborated process, but the conclusion is analogous: the PCR vector of regression coefficients has almost only two important contributions, centered at ca. 1900 and 2080 nm. Let us simplify matters, and consider negligible all $\mathbf{b}_n$ elements except those at 1900 and 2080 nm, which are approximately equal in magnitude and of opposite sign. The moisture level will be predicted simply by multiplying $\mathbf{b}_n$ by the sample spectrum $\mathbf{x}$. If only two elements of $\mathbf{b}_n$ are significantly different than zero, the product ($\mathbf{b}_n\,\mathbf{x}$) will only have two terms: ($k\,x_{1940}$) and ($-k\,x_{2080}$). The moisture level y will then be given by $y = k\,(x_{1940} - x_{2080})$, i.e., Norris equation!

5.8   A PCR Algorithm

A MATLAB algorithm for PCR can be written in only a few programming lines, as shown in Box 5.1. Is it too good to be true? Maybe: notice that 'princomp' in the first line of Box 5.1 invokes a MATLAB sub-routine which estimates the loadings and scores of the matrix 'X', and returns the variables 'U' and 'T'. If we could inspect this sub-routine in detail, the PCR algorithm would look considerably more complex than the three programming lines in Box 5.1.

Box 5.1

PCR algorithm: after the calibration data ('X' and 'yn'), the unknown sample spectrum ('x') and the number of latent variables 'A' are present in the workspace, the following codes provide the analyte concentration in the unknown ('y'):

> [U,T]=princomp(X','econ') ;
> bn=U(:,1:A)*inv(T(:,1:A)'*T(:,1:A))*T(:,1:A)'*yn;
> y=bn'*x;

The sizes of the input variables are: 'X', $J \times I$ (J = number of wavelengths or sensors, I = number of calibration samples), 'yn', $I \times 1$, 'x', $J \times 1$, and 'A', 1×1. Those generated during program execution are 'U', $J \times A$, 'T', $I \times A$, 'bn', $J \times 1$, 'y', 1×1.

5.9   How Many Latent Variables?

Until now we have mentioned the use of truncated matrices for calibration and prediction with the PCR model, but we have not explained in detail how to estimate the value of A. The next chapter will be completely devoted to this important subject. There are various mathematical techniques to estimate A, although there is some consensus in that one of them is to be preferred. Why would two procedures differ in the estimation of A? The answer is that the loadings, which we have classified in highly significant and poorly significant, should be really classified into three groups: highly significant, *doubtful,* and poorly significant. The region of doubtful loadings is the one that would make a statistical technique to say that A is 4, while another one says that it is 5. We may not expect that two different mathematical procedures to estimate A differ in 10 latent variables, but we may naturally expect differences in a few latent variables, depending on the specific details of the calculations.

5.10   Advantages of PCR

The main advantages of PCR are easy to summarize: (1) use of full-spectral data, (2) inverse calibration model, (3) replacement of the original signals, which are likely to be highly correlated, by latent variables with no correlation (orthogonal). There are additional advantages. As any model using full-spectral data, PCR should be able to provide spectral residuals for each unknown sample, restoring the first-order advantage presented by CLS, which was somewhat lost in ILS.

In the case of PCR, the spectral residuals are defined as the difference between the experimental sample spectrum (**x**) and the spectrum that can be reconstructed by the model ($\mathbf{x}_A$). This latter spectrum is found from the score vector for the sample and the truncated loading matrix $\mathbf{U}_A$:

$$\mathbf{x}_A = \mathbf{U}_A \mathbf{t}_A \tag{5.11}$$

This latter expression can be interpreted by saying that the estimated spectrum is the reconstruction or decompression of the sample scores, from the latent space to the real space. As always, the link between both spaces is the truncated loading matrix.

We could define a parameter analogous to that employed in CLS as a measure of the spectral residue:

$$s_{\text{res}} = \sqrt{\frac{\sum\limits_{j=1}^{J} \left(x_j - x_{Aj}\right)^2}{J - A}} \tag{5.12}$$

where x_j and x_{Aj} are generic elements of the sample spectrum $\mathbf{x}$ and of the reconstructed spectrum $\mathbf{x}_A$ with A latent variables as shown in Eq. (5.11). In a future chapter we will provide a more elaborated test for judging whether the spectral residues for a given sample are significant or not.

In this way, the first-order advantage is added to the list of advantages already given for PCR, because s_{res} will be comparable to the instrumental noise level for *normal* samples, those not carrying unexpected constituents. Conversely, s_{res} will be significantly larger than the instrumental noise, flagging samples with uncalibrated constituents as abnormal. As was the case in CLS, the spectrum of residuals ($\mathbf{x} - \mathbf{x}_A$) will show random noise in the absence of unexpected interferents, and spectral features otherwise.

A final characteristic of PCR is that the scores of the calibration samples, and also those for validation and true unknowns, can be placed in a score–score plot (second score vs. first score), allowing one to get a rapid visualization of the mutual relationship among samples. For example, we could discover in this plot that samples that were not used to build the model (either validation or unknowns) are outside the region delimited by the calibration space. If these samples are validation ones, they should be included (along with other ones of similar behavior) in the calibration set, to provide more representativity to the latter. If they are true unknowns, it is likely that they contain new constituents, and could not be analyzed with the current model.

5.11 A Real Case

A paradigmatic example of the successful application of PCR to a real analytical problem is the determination of parameters of interest in whole seed samples by means of NIR spectroscopy, replacing the classical analyses of oil, moisture, protein,

Table 5.1 Statistical indicators for the PCR validation phase in the analysis of seeds[a]

Parameter	RMSEP/%	REP/%
Moisture	0.016	0.16
Total oil	0.08	2.0
Total protein	0.11	1.2
Starch	0.17	0.3

[a]*RMSEP* root mean square error in prediction (%), *REP* relative error of prediction (%)

Fig. 5.6 NIR spectra of corn seeds, employed to validate a PCR model for the rapid determination of quality parameters. See caption of Fig. 5.4

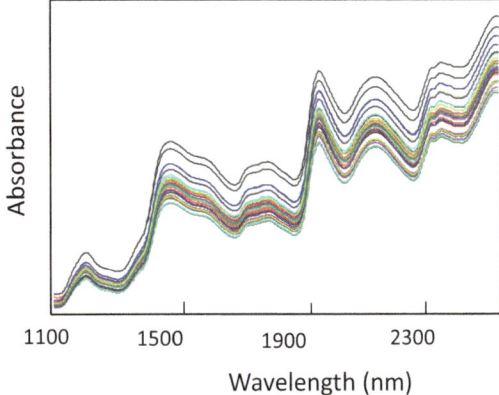

and starch. Some of them employ toxic organic solvents, and are expensive, complex, and time consuming (moisture evaporation, oil extraction, protein digestion, and starch hydrolysis take hours). The combination NIR/PCR, instead, allows one to perform the simultaneous analysis of the four parameters in a short time, with no use of solvents or auxiliary reagents, with a low cost per sample and without processing the seed samples (in the extreme case, without even grinding the seeds). This type of analysis is today routine in quality control laboratories of agricultural products around the world.

Figure 5.4 shows the NIR spectra of a set of 50 corn seed samples,[1] in the wavelength range from 1100 nm to 2498 nm each 2 nm (700 different wavelengths). Notice the high degree of spectral overlapping or correlation among the NIR profiles in Fig. 5.4. Without PCR, the probability of success would be low.

The four parameters of interest in the seeds employed for calibration were independently measured by reference analytical techniques. The experimental ranges for these properties are: moisture, 9.38–10.99%; total oil, 3.09–3.83%; total protein, 7.65–9.71%; and starch, 66.47–62.83%. The aim is to build PCR models to replace the classical determinations. To illustrate the result, we show in Table 5.1 the average prediction error in these parameters, estimated for an independent group of 30 validation samples, employing full-spectral PCR models with 20 latent variables ($A = 20$). The NIR spectra of the validation samples are shown in Fig. 5.6. We can

---

[1]This data set is freely available in the internet at: http://www.eigenvector.com/data/Corn/.

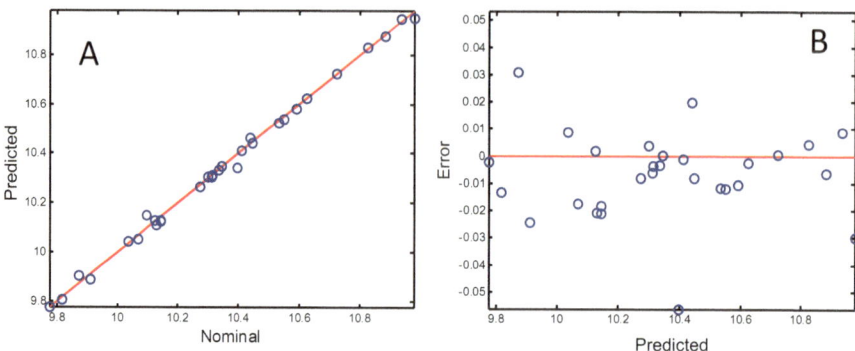

Fig. 5.7 (a) Predicted moisture values as a function of nominal ones for the samples of Fig. 5.6. (b) Prediction errors as a function of predicted values. In (a) the red line indicates the perfect correlation of unit slope; in (b) the red line indicates null errors

judge the quality of the calibration with the aid of the average errors, both absolute (RMSEP) and relative (REP) as shown in Table 5.1.

The results are encouraging. They imply very good statistical parameters for the analysis of all four properties. However, *very good* is a qualitative observation, and needs to be further substantiated. How good are the REP values for the validation? A rule of thumb in validation, already advanced in Chap. 1, is the *2-5-10* rule: less than 2%, excellent; less than 5%, reasonably good; less than 10%, good; more than 10%, poor. In any case, a relative error needs to be compared with the one for the reference technique, or with the tolerable error according to the international protocols.

In the case of Table 5.1, the relative errors should be compared with the typical errors involved in the corresponding reference techniques. The literature indicates that the repeatability of the classical determinations are ca. 0.15% for moisture (ISO 6540 1980), 2.5% for total oil (ISO 6492 1999), 1.5% for protein (ISO 20483 2000), and 1.5% for starch (ISO 6493 2000). The relative errors obtained by NIR/PCR (Table 5.1) are comparable, with a considerably saving of time (only the extraction phase of the classical methods takes hours), with no use of toxic organic solvents or other reagents, and providing, at the same time, the values of additional sample properties.

Figures 5.7 and 5.8 show more information on the validation phase: the plots of predicted vs. nominal property values and prediction errors vs. predicted values for the parameters moisture and total oil, respectively. The one for moisture is excellent (Fig. 5.7); we expect this result because NIR spectroscopy is highly sensitive to the vibrations of the water molecule. The plot for oil (Fig. 5.8) is less impressive, but still satisfactory for routine quantitative analysis.

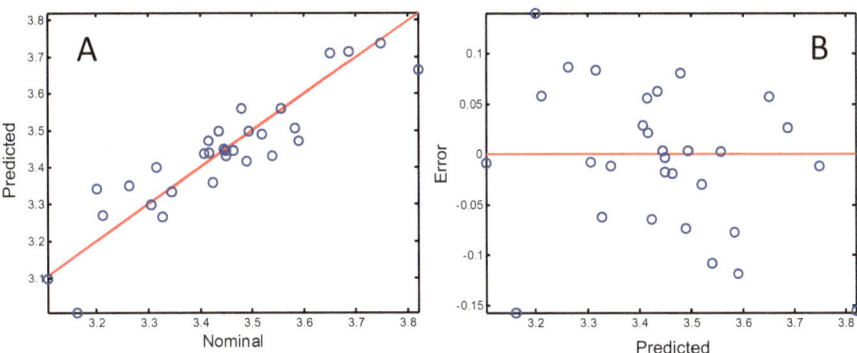

Fig. 5.8 (**a**) Predicted total fat values as a function of nominal ones for the samples of Fig. 5.6. (**b**) Prediction errors as a function of predicted values. In (**a**) the red line indicates the perfect correlation of unit slope; in (**b**) the red line indicates null errors

5.12 What Can Be Better Than PCR?

If PCR displays all the advantages of CLS and ILS, and none of their limitations, the logical question is: who can overcome PCR? The answer is not simple. The general consensus appears to be that there is room for improvement. A defect that can be noticed in PCR is that the latent variables are estimated using only the spectral information for the calibration set, without using information on analyte concentrations or sample properties which is available for the calibration samples. This latter information is valuable, and multivariate models based on the combination of spectral and concentration data for computing the latent variables could be able to improve the predictive ability of PCR.

In a future chapter we will discuss a model that employs all the available information when estimating the latent variables: partial least-squares (PLS) regression (Wold et al. 2001). The PLS model is today the de facto standard for first-order multivariate calibration, and the *official* explanation for this fact is the above one. However, not all authors agree with this view (Wentzell and Vega Montoto 2003; Lin et al. 2016). The apparent truth, almost universally accepted, that PLS is better than PCR is being challenged, and there are researchers thinking that deep down PCR and PLS are equivalent (Wentzell and Vega Montoto 2003; Lin et al. 2016).

5.13 Exercises

1. When trying various multivariate models for predicting the content of foreign oils and fats in butter fats from gas chromatographic data (Lipp 1996), an ILS model was found to be appropriate for the detection of the addition of about 3–5% foreign fat, whereas PCR calibration leads to a model with 11 latent variables

indicating a detection limit of about 3%. Can you give an explanation for the better PCR results in comparison with ILS?

2. In a discussion forum on multivariate models, a researcher once asked: when is it necessary to move from traditional ILS to PCR? Two rather opposite answers were given to this question.

   (a) Always. ILS is, after all, just a special case of PCR. If you include all the components, you arrive at the same model that ILS gives you. But along the way you get additional diagnostic information from which you might learn something. The scores and loadings all give you information about the chemistry.

   (b) As a grumpy old man, I say the time to switch to PCR is when you are ready to admit that you don't have the knowledge or patience to do actual spectroscopy.

What do you think?

References

ISO 20483: Cereals and Pulses—Determination of the Nitrogen Content and Calculation of the Crude Protein Content—Kjeldahl Method. International Organization for Standardization (ISO), Geneva (2000)

ISO 6492: Animal Feeding Stuffs—Determination of Fat Content. International Organization for Standardization (ISO), Geneva (1999)

ISO 6493: Animal Feeding Stuffs—Determination of Starch Content—Polarimetric Method. International Organization for Standardization (ISO), Geneva (2000)

ISO 6540: Maize—Determination of Moisture Content (on Milled Grains on Whole Grains). International Organization for Standardization (ISO), Geneva (1980)

Lin, Y.W., Deng, B.C., Xua, Q.S., Yun, Y.H., Liang, Y.Z.: The equivalence of partial least squares and principal component regression in the sufficient dimension reduction framework. Chemom. Intell. Lab. Syst. **150**, 58–64 (2016)

Lipp, M.: Comparison of PLS, PCR and MLR for the quantitative determination of foreign oils and fats in butter fats of several European countries by their triglyceride composition. Z. Lebensm. Unter. For. **202**, 193–198 (1996)

Norris, K.H., Hart, J.R.: Direct spectrophotometric determination of moisture content of grain and seeds. In: Principles and methods of measuring moisture in liquids and solids. Proceedings of the 1963 International Symposium on Humidity and Moisture, vol. 4, pp. 19–25. Reinhold, New York (1965)

Thomas, E.V., Haaland, D.M.: Partial least-squares methods for spectral analyses. 1. Relation to other quantitative calibration methods and the extraction of qualitative information. Anal. Chem. **60**, 1193–1202 (1988)

Wentzell, P.D., Vega Montoto, L.: Comparison of principal components regression and partial least squares regression through generic simulations of complex mixtures. Chemom. Intell. Lab. Syst. **65**, 257–279 (2003)

Wold, S., Sjöström, M., Eriksson, L.: PLS-regression: a basic tool of chemometrics. Chemom. Intell. Lab. Syst. **58**, 109–130 (2001)

Abstract

The relevant issue of optimizing the number of latent variables in full-spectral inverse models is discussed, with emphasis on interpretation rather on statistical and mathematical issues.

6.1 How Many Latent Variables?

We have remarked several times the importance of estimating an optimum value of A, the number of latent variables for truncating the score matrix and building a PCR model, without additional details. We now give the reasons why this activity is crucial for the model success.

As an analogy, we first explore a simpler problem. Suppose we have a series of experimental values of the variable y, measured at different values of the variable x in the range 0–5, as shown in Fig. 6.1, and we wish to predict the value of y for $x = 6$.

To answer this question, we should first model the y data in the range of x values from 0 to 5 through a suitable mathematical expression. Three different polynomial equations will be used for this purpose: (1) a linear model ($y = a + bx$), (2) a quadratic model ($y = a + bx + cx^2$), and (3) a cubic model ($y = a + bx + cx^2 + dx^3$). In each case, the adjustable model parameters are: (1) a and b; (2) a, b, and c; and (3) a, b, c, and d. We may say that the model complexity grows from (1) to (3), while the number of adjustable parameters increases. Deep down, we are exploring the following question: is it worth to use a more complex model to fit the data?

Figure 6.2 shows the straight line fitted according to the linear model, and the resulting sum of squared residuals (SSR). This indicator is the sum of the square differences between experimental values of y and the values predicted by the fitted model. The lack of fit is apparent, since the value of SSR is large; the square root of this number, adjusted by the number of observations (11), is ca. 4.8 units. This is

© Springer Nature Switzerland AG 2018

A. C. Olivieri, *Introduction to Multivariate Calibration*,

https://doi.org/10.1007/978-3-319-97097-4_6

Fig. 6.1 Plot of values of y as a function of x for an experimental data set (circles). The red circle marks the point where extrapolation is required

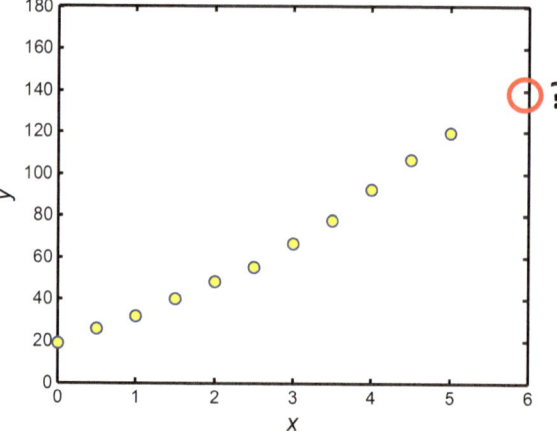

Fig. 6.2 Regression (green) line fitted to the data in Fig. 6.1 using the linear model. The value of SSR is the sum of squares of model residues

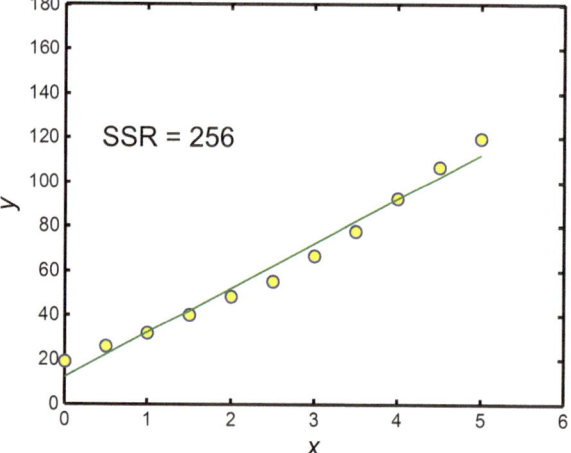

close to 7% of the average of the measured y values, and we expect a considerably better fit with more complex models. In addition, the residuals (differences between predicted and experimental values) show a clear correlation: several positive residuals occur in the center of the x range, and negative ones in the extremes (Fig. 6.2).

If the polynomial degree is increased to 2, moving to the quadratic model, a significantly better fit is obtained (Fig. 6.3): the value of SSR considerably decreases. Adjusting SSR by the number of observations and taking the square root leads to a lower typical error of 1.4%. On the other hand, at least visually, no significant correlation appears to exist among the residuals.

What happens if we further increase the degree of the polynomial model? The fit to the cubic model improves (Fig. 6.4) judging from the new value of SSR, with a typical error of 1.36%. Is this improvement significant or only marginal? At first

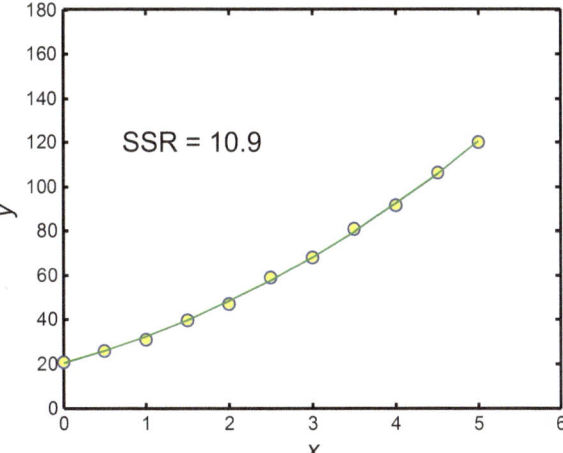

Fig. 6.3 Regression (green) line fitted to the data in Fig. 6.1 using the quadratic model. The value of SSR is the sum of squares of model residues

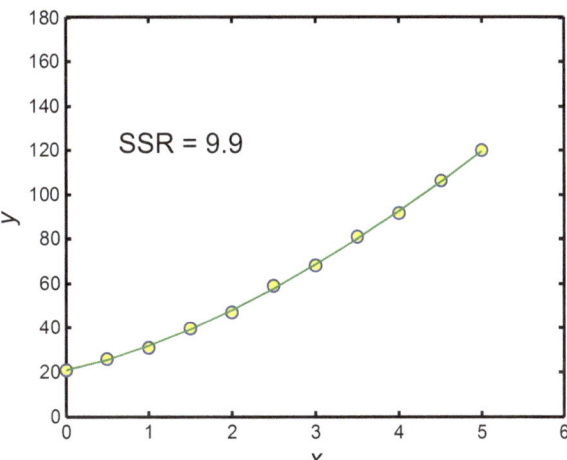

Fig. 6.4 Regression (green) line fitted to the data in Fig. 6.1 using the cubic model. The value of SSR is the sum of squares of model residues

sight, both the decrease in SSR and typical error appear to be small and marginal, but it is a fact that the error decreases. What would go wrong if we use the most complex cubic model?

Going back to the original aim, to estimate the value of y for $x = 6$, the question is whether it is best to do it with the quadratic or with the cubic model. Figure 6.5 shows this extrapolating activity with the quadratic model. We call it extrapolation since we intend to estimate y for an x value outside the calibration range of the model. As can be seen, interpolation within the calibration range is reasonably safe. Extrapolation at $x = 6$ is also safe, although with a somewhat larger uncertainty.

On the other hand, when the cubic model is employed, although interpolation is safe, extrapolation is considerably more uncertain than for the quadratic model (Fig. 6.6). We may summarize these results by saying that the linear model under-fits the data, with less adjustable parameters than those really needed. The quadratic

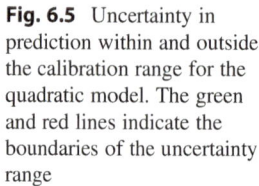

Fig. 6.5 Uncertainty in prediction within and outside the calibration range for the quadratic model. The green and red lines indicate the boundaries of the uncertainty range

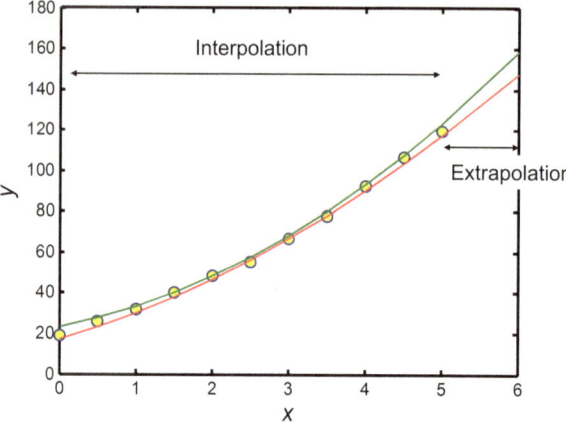

Fig. 6.6 Uncertainty in prediction within and outside the calibration range for the cubic model. The green and red lines indicate the boundaries of the uncertainty range

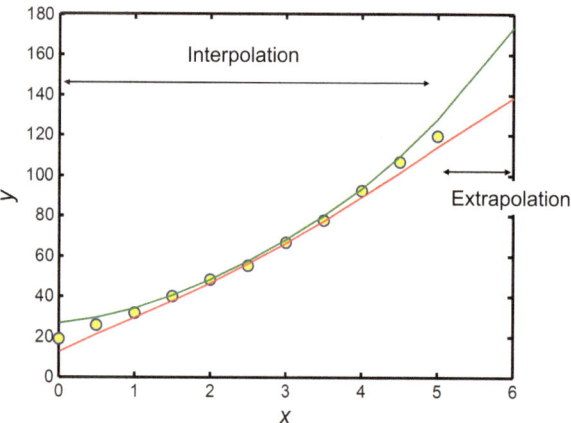

model yields a reasonable fit. Finally, the cubic model represents an over-fitting of the data, with more adjustable parameters than needed, and more uncertainty for extrapolation.

Figure 6.7 graphically collects this information: with fewer adjustable parameters, the model lacks predictive power for future data points not included in the calibration, leading to a large bias in prediction. Increasing the number of parameters improves the predictive power, but the prediction variance gradually increases, in such a way that a certain optimum number of parameters occurs, which may ensure reasonable prediction in terms of both bias and variance. The overall situation is known as the *bias–variance compromise*.

We can easily extend the analogy to PCR calibration. If a small number of latent variables is employed for building the calibration model, it carries the risk of predicting the analyte concentrations in future samples with a considerable bias, because the model lacks sufficient information. If we exaggerate in the number of latent variables, some of them will be modeling the instrumental noise. However,

Fig. 6.7 Illustration of the
bias–variance compromise

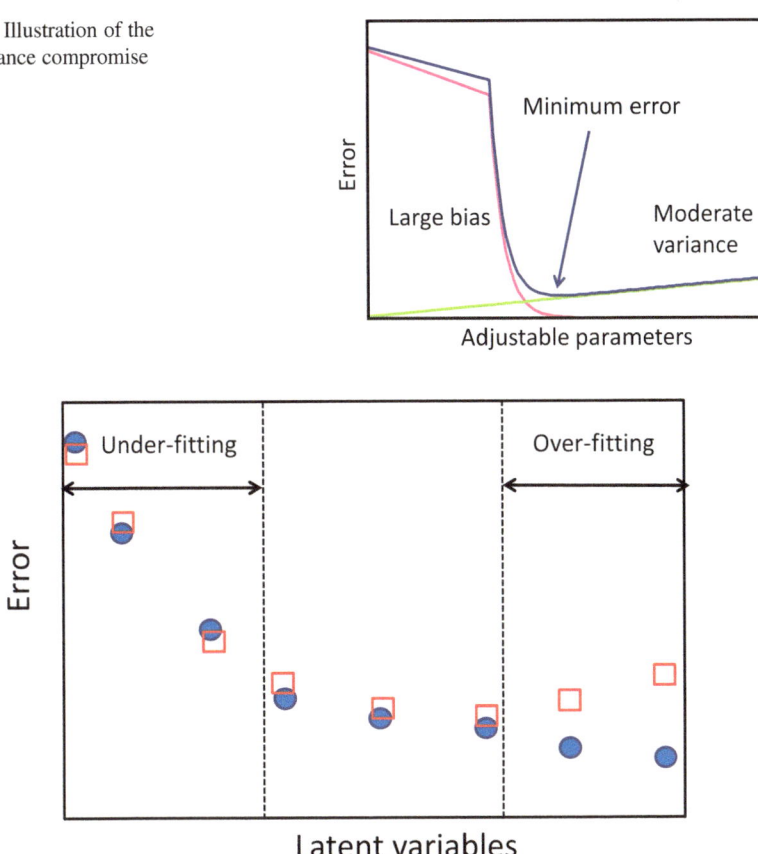

Fig. 6.8 Changes in prediction error for calibration samples (blue circles) and for independent samples (red squares) as a function of the number of latent variables

due to its random nature, the noise in the new samples is different than the noise in the calibration samples, so that the model will not be able to account for it. The result will be a prediction of lower quality and larger uncertainty. As a consequence, the bias–variance compromise implies that the selection of the optimum value of A will be of great importance to build a successful PCR model.

Figure 6.8 shows the typical progression of the prediction error as a function of A for the calibration and for independent samples. When estimating the analyte concentrations in the same samples used to build the model, the error continuously decreases. This is a rather false feeling, because the latent variables representing instrumental noise are being employed to study samples with the same noise structure. When independent samples are studied, the prediction error decreases at first, as more valuable information is included in the model, until over-fitting is reached, where the error begins to increase.

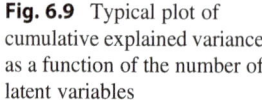

Fig. 6.9 Typical plot of cumulative explained variance as a function of the number of latent variables

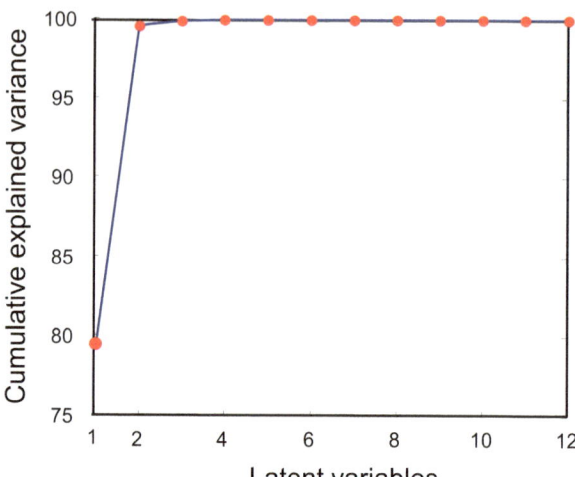

In addition, the number of latent variables may affect the model robustness over time. Too many latent variables tend to be more sensitive to minor concentrations of new constituents in test samples. In contrast, models with fewer latent variables do not significantly degrade its predictive power over time.

6.2   Explained Variance

We have already explored the concept of explained variance in Chap. 4. One alternative to set the optimum number of latent variables is to consider the cumulative explained variance by the successive principal components, starting from the most contributing one, until a certain % of the total variance is reached.

Figure 6.9 illustrates a typical behavior for a set of principal components of an experimental system: the individual contribution of each component progressively decreases, while the cumulative explained variance increases. In this case, for example, the first three principal components justify more than 99% of the spectral variance, a fact that may lead to the conclusion that the recommended value of A is 3 for truncating the score matrix.

This method of estimating the value of A has two problems. First, different authors employ different criteria for the optimum % of cumulative variance. On the other hand, only spectral data are used to estimate A, leaving the analyte concentration information in the calibration samples unused. In analytical chemistry, it is preferable to incorporate in this analysis the available information regarding the calibration concentrations, together with the calibration signals.

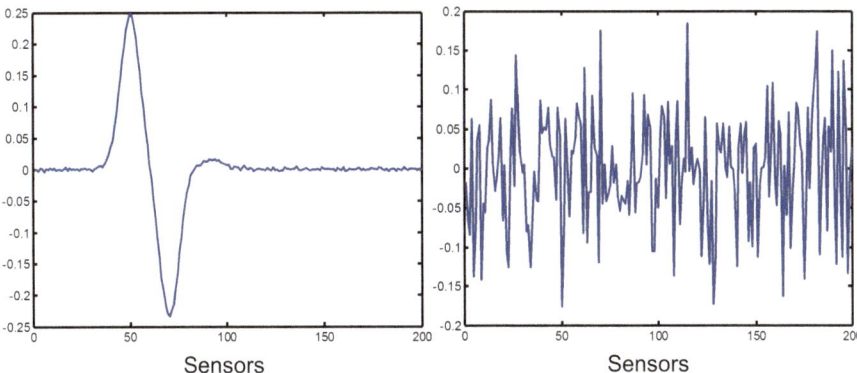

Fig. 6.10 Left, a highly significant loading representing instrumental variance due to physicochemical phenomena. Right, a poorly significant loading representing instrumental noise

6.3 Visual Inspection of Loadings

Beyond the mathematical and statistical tools to estimate the value of A, visual inspection of the loadings is important. Figure 6.10 shows the difference between a loading representing real physicochemical phenomena, which can be considered as highly significant, and a loading basically representing noise (a poorly significant one).

Figure 6.10 shows two cases than can be considered extreme. In real life the situation can be more complex, and it might be difficult to discern whether a loading is highly or poorly significant based on visual inspection. This is especially so for the intermediate loadings, which contain both useful information and instrumental noise in comparable degrees. Figure 6.11 shows the first six loadings, as estimated with PCA, for a real analytical system: the determination of an active principle in a cough syrup, where the analyte and the remaining syrup constituents show a high degree of spectral overlapping in the UV-visible region. The first three loadings have a clear spectral aspect, but from the fourth and beyond the classification into highly or poorly significant becomes difficult.

To avoid human-dependent influences in cases such as the above one, more objective mathematical/statistical strategies have been designed. The most employed one in multivariate calibration is described in the next section.

6.4 Leave-One-Out Cross Validation

The possibility of calibrating and predicting by means of a PCR model offers an interesting alternative for the selection of the optimum number of latent variables A through a combination of spectral information and nominal analyte concentrations

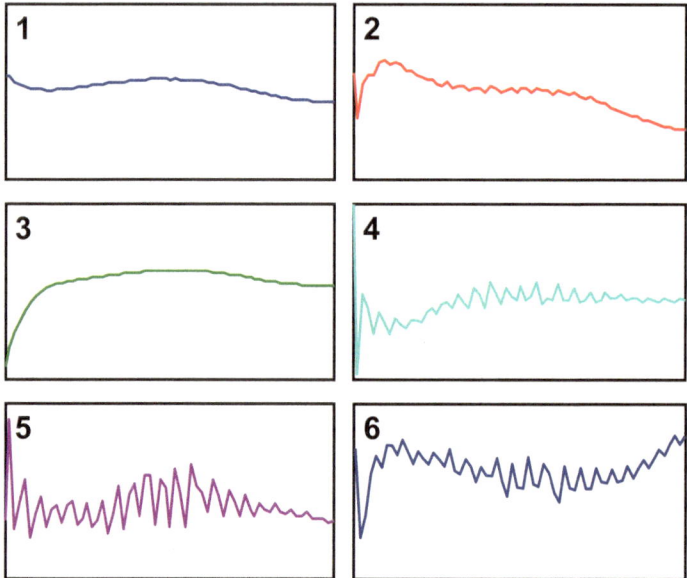

Fig. 6.11 Plot of the six first loadings of an experimental system studied by PCA

(or sample properties). The procedure is known as cross validation (Thomas and Haaland 1988). It consists in calibrating various PCR models with sub-sets of samples taken from the calibration set, leaving the remaining ones for prediction. When a single sample is left out, the methodology is known as leave-one-out (LOO) cross validation.

In the LOO procedure, the analyte concentration in the left out sample is estimated with a model built by the remaining samples, and the prediction error is computed (difference between nominal and predicted value), in principle using a single latent variable ($A = 1$). The left out sample is then returned to the calibration set, and samples are individually left out, until every sample has been left out once. Prediction errors using a single latent variable are computed in each case. The procedure is repeated using an increasing number of latent variables until a certain maximum value A_{max}. For each number of latent variables, the following parameter is calculated: the prediction error sum of squares (PRESS) for all left out samples. Then the changes in PRESS values as a function of the number of latent variables are analyzed with a statistical method. Table 6.1 summarizes the steps to be followed for LOO cross validation.

As an example, during a typical experimental LOO cross validation procedure, Table 6.2 was obtained showing PRESS values as a function of the number of latent variables up to $A_{max} = 6$. A convenient plot of the PRESS values is shown in Fig. 6.12.

Table 6.2 and Fig. 6.12 show that, as latent variables are added to the model (from $A = 1$ to $A = 4$), the PRESS decreases. This is due to the fact that the data

Table 6.1 Steps in LOO cross validation

Step	Activity
1	A sample is left out from the calibration set
2	A PCR model is built with one latent variable and the remaining samples
3	The analyte concentration in the left out sample is estimated by the PCR model (y_{pred})
4	The quadratic prediction error is calculated, as the square of the difference between predicted and nominal value ($y_{pred} - y_{nom})^2$
5	The left out sample is returned to the calibration set, and another sample is left out
6	The process is repeated from step 1 to step 5, until all samples have been left out once
7	The PRESS $= \Sigma(y_{pred} - y_{nom})^2$ is calculated (the summation has as many terms as calibration samples)
8	The process is repeated from step 1 to step 6, using 2, 3, ..., A_{max} latent variables
9	A list of PRESS values is obtained as a function of the number of latent variables

Table 6.2 PRESS values as a function of the number of latent variables in an experimental system

Latent variables	PRESS
1	0.60
2	0.0120
3	2.8×10^{-3}
4	2.1×10^{-3}
5	3.6×10^{-3}
6	0.0114

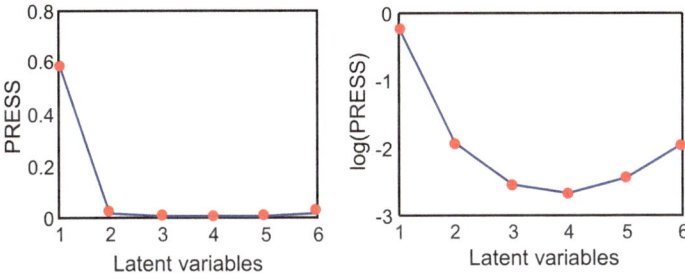

Fig. 6.12 Changes in PRESS (left) and log(PRESS) (right) as a function of the number of latent variables for a typical PCR model

compression is progressively more efficient; the first latent variables contain relevant information regarding the spectral variations in the calibration set. Further increasing the number of latent variables (from $A = 4$ to $A = 6$) causes the PRESS to slightly increase. This is a strong indication that the last latent variables are not providing relevant information, and that they essentially represent noise. Using more latent variables may lead to the undesired situation of over-fitting. Why does the PRESS increase from a certain value of A? In cross validation, the left out samples not included in the sub-set employed for calibration act as independent samples, different than those used for model building. This situation is similar to the one shown in

Fig. 6.8, where the prediction error in new samples increases when the number of latent variables is too large.

An important question arising in connection with the above process is: which value of A_{max} should be selected for a given LOO cross validation study? Some useful suggestions follow. First, the optimum value of A cannot exceed the number of calibration samples, so that A_{max} should be smaller than I. Some generosity should accompany this criterion, i.e., A_{max} cannot be $(I-1)$ or so. It has been proposed, for example, that A_{max} should be equal to half the number of calibration samples $(I/2)$, although no firm theoretical justification exists for this criterion. Another useful suggestion is that the PRESS plot as a function of the number of latent variables should show a minimum. If a certain value of A_{max} was selected, and the minimum does not appear, one should increase A_{max} until observing the minimum. However, if this value approaches the number of samples, it may indicate that you need more calibration samples!

In the present context, it is important to notice that each PRESS value is linked to a certain value of the average cross validation error $(PRESS/I)^{1/2}$, which is called the root mean square error in cross validation (RMSECV). The RMSECV value for the minimum PRESS should be consistent with the expected uncertainty in prediction for future samples, or with the uncertainty associated to the nominal concentration values or sample properties used for model building. Otherwise, although a minimum PRESS is obtained at a certain value of A, the corresponding calibration error may not be acceptable.

6.5   Cross Validation Statistics

Intuitively, one may select the optimum number of latent variables as the one leading to the minimum PRESS. However, statistics indicates that this might not be the wisest selection. A convenient technique to find the optimum value of A is the one described by Haaland and Thomas (Thomas and Haaland 1988). It consists in expanding Table 6.2 with additional columns, computing the ratios between the PRESS values and the minimum one (only for fewer latent variables than those producing the minimum PRESS). These PRESS ratios play the role of variance ratios (analogous to the statistical F parameter), so that they can be associated to a

Table 6.3 Statistical analysis of LOO cross validation results[a]

Latent variables	PRESS	PRESS/min(PRESS)	p	RMSECV
1	0.60	280	0.999	0.22
2	0.012	5.62	0.997	0.032
3	$\mathbf{2.8 \times 10^{-3}}$	**1.32**	**0.679**	**0.015**
4	2.1×10^{-3}	1	0.5	0.013
5	3.6×10^{-3}	–	–	–
6	0.011	–	–	–

[a]Optimum values in boldface

probability p, estimated with a number of degrees of freedom equal to the number of calibration mixtures I both for the numerator and denominator. The results from Table 6.2 are shown in Table 6.3.

The proposal, based on empirical results, is that the number of latent variables to be selected is the first one in Table 6.3 for which the associated probability p is lower than 0.75 (Thomas and Haaland 1988). This criterion leads to $A = 3$ as the optimum value for building the PCR model, although the PRESS is minimum at $A = 4$. The interpretation here is that the PRESS value for three latent variables is not statistically larger than the PRESS value for four latent variables. The recommendation is to select the smallest number of latent variables for which the PRESS is not statistically different than the minimum, in this case, three.

The optimum PRESS and its associated RMSECV value can be employed to gather an idea of the predictive ability of the model during the cross validation phase. The RMSECV should be of the order of the uncertainty associated to the calibration concentrations. In the last column of Table 6.3, we see that for $A = 3$, the RMSECV is 0.015 units. This number should have a qualitative value, judiciously employed by the analyst. In the present experimental case, the calibration concentrations were in the range from 1.55 to 2.66 (in units of 10^{-4} mol L^{-1}) with an average of 2.06 (Goicoechea and Olivieri 1999). The uncertainty for preparing these concentrations can be estimated as ca. 0.01–0.02 units, so that a cross validation error (RMSECV) of 0.015 units is highly reasonable.

6.6   Monte Carlo Cross Validation

When the number of calibration samples is large, as is usual in the study of complex samples with a large number of variability sources, LOO cross validation may be time consuming. In these cases one may employ an alternative procedure: instead of leaving a single sample out of the calibration set, a sub-set of samples is left out. The proportion of left out samples varies, usually 30% of the total number of samples are left out, using the remaining 70% for model building at each stage of the process. These calculations are repeated a number of times (e.g., 10 times), selecting the left out samples at random each time. The final result is a table of PRESS values as a function of the number of latent variables, analogous to Table 6.3, whose statistical analysis proceeds in a similar manner as discussed above. This procedure is called Monte Carlo cross validation, because of the introduction of random elements in its formulation (Xua and Liang 2001). Table 6.4 summarizes the steps to be followed in this case.

6.7   Other Methods

Cross validation is the most employed methodology to assess the optimum number of latent variables in multivariate models such as PCR. However, a number of drawbacks have been discussed by some authors, as listed below.

Table 6.4 Steps in Monte Carlo cross validation

Step	Activity
1	A sub-set of samples (30% of the total number of calibration samples, selected at random) is left out from the calibration set
2	A PCR model is built with a single latent variable with the remaining 70% samples
3	The analyte concentration or sample property is estimated in the left out sub-set of samples (one y_{pred} value per sample)
4	The quadratic prediction error is calculated as the sum of squares of differences between predicted and nominal values $\sum(y_{pred} - y_{nom})^2$
5	The left out sub-set of samples is returned to the calibration set, and another random sub-set of samples is left out
6	The process is repeated from step 1 to step 5 a number of times (e.g., 10 times)
7	The PRESS is calculated as $\sum\sum(y_{pred} - y_{nom})^2$
8	The process is repeated from step 1 to step 7 using 2, 3, ..., A_{max} latent variables
9	A list of PRESS values is obtained as a function of the number of latent variables

1. The values of PRESS are reliable if the reference values used for calibration (analyte concentration or sample properties) are known with sufficient accuracy. Otherwise, the calibration errors may obscure the true changes in PRESS which are required to estimate the optimum A.
2. The plot of PRESS as a function of latent variables exhibits a relatively flat region around the minimum, making it difficult to identify a reliable value for the optimum A which does not significantly differ from the minimum.
3. The procedure is not recommended when the calibration set is statistically designed. Removing samples of a design for building a model with the remaining samples leads a loss of the design properties.

To avoid these inconveniences, a method has been proposed based on the analysis of subsequent latent variables added to the model, using randomization procedures (Faber and Rajko 2007). Briefly, this method randomly permutes the values of analyte concentrations, in such a way that any relationship between the instrumental data matrix $\mathbf{X}$ and the target concentration vector $\mathbf{y}_n$ is destroyed. If the latent variables reflect real phenomena occurring in the calibration set, permutation will generate PCR models which will be significantly different than the original model. Conversely, a poorly significant latent variable will not allow for distinguishing permuted and non-permuted data. The report describing this method provides indications for interpreting the results (Faber and Rajko 2007).

6.8   The Principle of Parsimony

There are philosophical reasons for not abusing the number of latent variables. One is the so-called principle of parsimony, announced centuries ago by William of Ockham (ca. 1286–1347): *frustra fit per plura quod potest fieri per pauciora*, translated as *it is useless to do with more what can be done with less*. Also called

Ockham's razor, it advises analysts that it is always preferable to use the most conservative number of latent variables. If one has to decide between two alternative models, which do not significantly differ in predictive ability, the principle indicates that one should select the most parsimonious one, i.e., the one involving the smallest number of adjustable parameters, or latent variables in PCR. In summary, the simplest one.

Another reason is the interpretability of A. From a qualitative point of view, the latent variables represent the sources of spectral variance in the calibration set, or the sources of variation of the measured instrumental signal from sample to sample. Therefore, the optimum number of latent variables should represent, in principle, the number of analytes generating a measurable signal. It may be larger than the actual number of analytes if other phenomena producing spectral changes occur, such as dispersion of radiation, baseline drifts, non-linearities, etc., but cannot be, in general, significantly larger.

The analyst should always take care that black-box procedures, such as cross validation, do not lead to an optimum number of latent variables which does not have any connection with the real world.

6.9   A Real Case

We now describe the multivariate spectral determination of fluoride in water samples. The standard spectrophotometric procedure for measuring fluoride is based on the effect of the latter on the zirconium complex of 2-(p-sulfophenylazo)-1,8-dihydroxy-3,6-naphthalene-disulfonate (SPADNS). Fluoride efficiently binds to zirconium, decreasing the absorbance of the complex, and the magnitude of the decrease is proportional to fluoride concentration (Method 4500 F D 1998). However, sulfate ions, which are regularly present in subterranean waters, are also able to react with zirconium, constituting an interference in the fluoride determination.

The interference can be modeled by multivariate calibration of visible spectra of a set of carefully designed samples (Arancibia et al. 2004). Figure 6.13 shows the spectra for 16 calibration samples, which are aqueous mixtures of fluoride, sulfate, and the zirconium complex (Arancibia et al. 2004). If the data are employed to build a PCR model, using the absorbances in the range 560–640 nm, the LOO cross validation results shown in Table 6.5 are obtained using $A_{max} = 8$. The statistical analysis discussed above leads to the conclusion that the optimum number of latent variables is four, despite the fact that a minimum PRESS is attained for five latent variables. The corresponding RMSECV value was found to be reasonably low, with additional details provided in (Arancibia et al. 2004) regarding validation and test samples.

What about the chemical interpretation of the optimum number of latent variables? Considering that fluoride and sulfate ions do not absorb in the working range of wavelengths, and that the only absorbing species appears to be the zirconium-SPADNS complex, one could reasonably ask: why four latent variables?

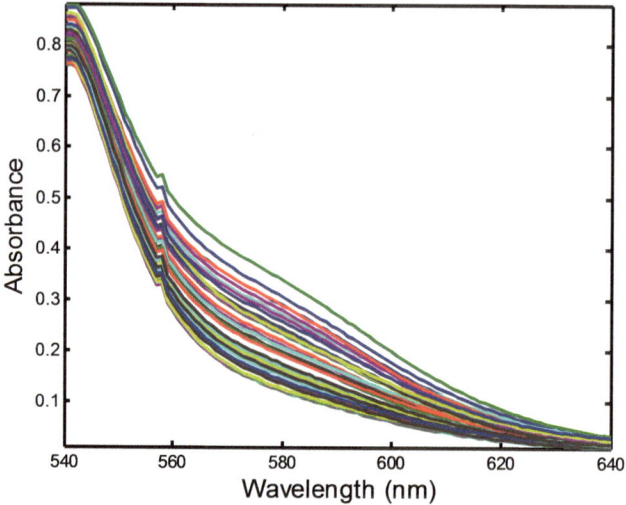

Fig. 6.13 Visible spectra of 16 aqueous mixtures of fluoride, sulfate, and a zirconium complex, used to build a multivariate calibration model for the determination of fluoride in the presence of sulfate in subterranean water samples. The spectral region from 560 to 640 nm was employed for calibration

Table 6.5 Statistical analysis of LOO cross validation results in the fluoride determination using visible spectrophotometry and multivariate calibration[a]

Latent variables	PRESS	PRESS/min(PRESS)	p	RMSECV
1	0.70	6.50	0.9990	0.14
2	0.23	2.16	0.9880	0.08
3	0.23	2.14	0.9870	0.08
4	**0.12**	**1.08**	**0.5890**	**0.06**
5	0.11	1	0.5	0.05
6	0.11	–	–	–
7	0.11	–	–	–
8	0.12	–	–	–

[a]Optimum values in boldface

This is a solution system, so that dispersion effects cannot account for the results. The answer lies in the known tendency of Zr ions to form *ternary* complexes with dyes and either fluoride or sulfate (Arancibia et al. 2004). Therefore, in addition to the Zr-SPADNS complex, the system involves ternary Zr-SPADNS-fluoride and Zr-SPADNS-sulfate complexes, whose spectra differ from the one for the binary complex. Three of the four latent variables can be explained by these species, which show apparently different spectra in the working region. Only one additional latent variable is required by PCR cross validation, which could be easily explained by a slight baseline distortion or other physical phenomena.

6.10 Exercises

1. Indicate whether the following statements are true or false:
   (a) A spectral system with two responsive analytes in solution may require 10 latent variables for model building with PCR
   (b) A spectral system with two responsive analytes in the solid state may require 10 latent variables for model building with PCR
   (c) If the dispersion of the radiation is mathematically removed from the data, the number of latent variables should decrease
   (d) A spectral system with three responsive analytes in solution may require a single latent variable for model building with PCR

References

Arancibia, J.A., Rullo, A., Olivieri, A.C., Di Nezio, S., Pistonesi, M., Lista, A., Fernández Band, B. S.: Fast spectrophotometric determination of fluoride in ground waters by flow injection using partial least-squares calibration. Anal. Chim. Acta. **512**, 157–163 (2004)

Faber, N.M., Rajko, R.: How to avoid over-fitting in multivariate calibration—the conventional validation approach and an alternative. Anal. Chim. Acta. **595**, 98–106 (2007)

Goicoechea, H.C., Olivieri, A.C.: Determination of bromhexine in cough–cold syrups by absorption spectrophotometry and multivariate calibration using partial least-squares and hybrid linear analyses. Application of a novel method of wavelength selection. Talanta. **49**, 793–800 (1999)

Method 4500 F D: Standard Methods for the Examination of Water and Wastewater, 20th edn, pp. 4–62. American Public Health Association, Washington, DC (1998)

Thomas, E.V., Haaland, D.M.: Partial least-squares methods for spectral analyses. 1. Relation to other quantitative calibration methods and the extraction of qualitative information. Anal. Chem. **60**, 1193–1202 (1988)

Xua, Q.S., Liang, Y.Z.: Monte Carlo cross validation. Chemom. Intell. Lab. Syst. **56**, 1–11 (2001)

The Partial Least-Squares Model

7

Abstract

The most popular first-order model based on partial least-squares is presented, and a range of applications are shown, from single and multiple analyte determinations to sample discrimination.

7.1 The PLS Philosophy

The first-order multivariate model known as partial least-squares (PLS) regression intends to improve the PCR model described in the previous chapter, by introducing the information regarding the calibration concentrations in the estimation of the latent variables. In the framework of PLS, two types of loadings exist: the *weight* loadings, contained in the matrix $\mathbf{W}$, and the loadings, contained in the matrix $\mathbf{P}$. The columns of $\mathbf{W}$ are orthonormal, whereas the columns of $\mathbf{P}$ are neither orthogonal nor normalized. The purpose of these two different types of loadings is to explain the maximum spectral variance in the original data matrix $\mathbf{X}$, and at the same time to explain the maximum correlation between $\mathbf{X}$ and the analyte concentration vector $\mathbf{y}_n$.

In this way, PLS employs latent variables which are analyte-dependent, while PCR employs latent variables which are independent on the analyte. The introduction of analyte concentration information in the PLS model makes the latter specific for each analyte. This is the main reason why PLS is thought to overcome PCR: the PLS model is adapted to the needs of each analyte, because its latent variables are estimated, in part, as a function of the analyte concentrations in the calibration samples.

© Springer Nature Switzerland AG 2018

A. C. Olivieri, *Introduction to Multivariate Calibration*,

https://doi.org/10.1007/978-3-319-97097-4_7

7.2   Calibration Phase

In the PLS calibration phase, as in PCR, an inverse model is built by correlating the analyte concentrations with sample scores. The mathematical relationship is formally identical to that for PCR; the difference lies in the manner in which the PLS scores are estimated, by including both spectral and concentration information.

Specifically, the PLS calibration phase consists in the following inverse model:

$$\mathbf{y}_n = \mathbf{T}_A\mathbf{v}_n + \mathbf{e} \tag{7.1}$$

where $\mathbf{v}_n$ (of size $A \times 1$) is the vector of PLS regression coefficients defined in the PLS latent space, $\mathbf{T}_A$ is the truncated score PLS matrix, and $\mathbf{e}$ is a vector collecting the concentration modeling errors. Although the symbols are identical to those in PCR, the specific values of the elements of $\mathbf{v}_n$ and $\mathbf{T}_A$ differ between PCR and PLS.

Finding $\mathbf{v}_n$ from Eq. (7.1) completes the calibration phase (see Sect. 5.3 in Chap. 5):

$$\mathbf{v}_n = \mathbf{T}_A{}^+\mathbf{y}_n \tag{7.2}$$

Recall that the PCR score matrix was the projection of the original data matrix $\mathbf{X}$ onto the space defined by the first A columns of the loading matrix $\mathbf{U}$, i.e., $\mathbf{T}_A = \mathbf{X}^T \mathbf{U}_A$. In PLS, on the other hand, both the loadings $\mathbf{P}$ and the weight loadings $\mathbf{W}$ participate in estimating the calibration score matrix. As in PCR, they are also truncated to the first A columns, which should retain the main portion of both the spectral and concentration variance. The specific expression for the PLS score matrix is:

$$\mathbf{T}_A = \mathbf{X}^T\mathbf{W}_A(\mathbf{P}_A{}^T\mathbf{W}_A)^{-1} \tag{7.3}$$

Although the matrix product $(\mathbf{P}_A{}^T \mathbf{W}_A)$ is not diagonal, no serious issues are found in the above equation for its inversion, due to the properties of the loadings. Finally, it is important to remark that the columns of $\mathbf{T}_A$ in Eq. (7.3) are orthogonal. This removes the correlation problems that may be present in the original matrix $\mathbf{X}$, in the same way as in PCR.

7.3   Mathematical Requirements and Latent Variables

The mathematical requirements for the calibration PLS phase are analogous to those for PCR. The matrix inversion involved in Eq. (7.2) was trivial in PCR, because $\mathbf{T}_A$ was composed of orthogonal columns. In PLS the score matrix $\mathbf{T}_A$ is different than the one for PCR, but the columns are still orthogonal, so that the inversion needed in Eq. (7.2) is also trivial.

On the other hand, in the PLS model the optimum number of latent variables A is required, as in PCR. The usual procedure for this purpose is cross validation, either leave-one-out of Monte Carlo, as previously described for PCR.

7.4   Prediction and Validation Phases

In the prediction phase, the regression coefficients are employed to estimate the analyte concentration in a future sample. A previous step is required, as in PCR, to find the test sample scores, which proceeds by means of the truncated loading matrices $\mathbf{W}_A$ and $\mathbf{P}_A$. The latter ones are of the same size as the truncated PCR loading matrix $\mathbf{U}_A$ ($J \times A$). The specific PLS expression for the test sample score vector is:

$$\mathbf{t}_A = (\mathbf{W}_A^{\mathrm{T}}\mathbf{P}_A)^{-1}\mathbf{W}_A^{\mathrm{T}}\mathbf{x} \qquad (7.4)$$

and the prediction equation is:

$$y = \mathbf{v}_n^{\mathrm{T}}\mathbf{t}_A \qquad (7.5)$$

where y is the predicted analyte concentration in the test sample, whose spectrum is the vector $\mathbf{x}$ of Eq. (7.4).

7.5   The Vector of Regression Coefficients

As in PCR, it is possible to decompress the vector of PLS latent regression coefficients $\mathbf{v}_n$ back to the real space. The truncated loading matrices will be useful for this purpose:

$$\mathbf{b}_n = \mathbf{W}_A(\mathbf{P}_A^{\mathrm{T}}\mathbf{W}_A)^{-1}\mathbf{v}_n \qquad (7.6)$$

This vector can then be employed to estimate the analyte concentration in a test sample:

$$y = \mathbf{b}_n^{\mathrm{T}}\mathbf{x} \qquad (7.7)$$

so that *the equation* can be found, in the same way as some commercial developers of multivariate calibrations do.

How different are the PCR and PLS loadings and regression vectors? We use for the comparison the simulated system of four constituents of Chap. 3 (cf. Fig. 3.6), composed of 20 calibration samples with constituent concentrations in the range from 0 to 1. The loadings and regression coefficients are estimated for a single analyte of interest using both multivariate models. The loadings are shown in Fig. 7.1 (the PCR matrix $\mathbf{U}_A$) and Figs. 7.2 and 7.3 (the PLS matrices $\mathbf{W}_A$ and $\mathbf{P}_A$, respectively).

As can be seen, the first PCR and PLS loadings ($\mathbf{u}_1$ and $\mathbf{p}_1$, respectively) are similar, but the subsequent ones are different. The differences arise from the specific manner in which the models estimate the latent variables. Table 7.1 lists the % of explained spectral variance by each latent variable in PCR and PLS. The values are

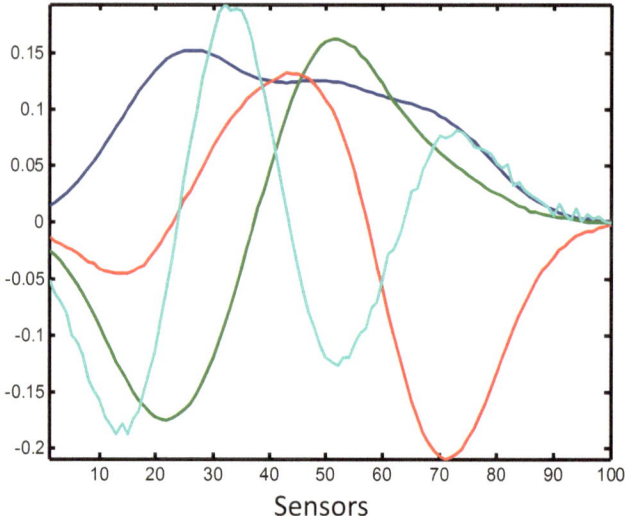

Fig. 7.1 First four PCR loadings (matrix $\mathbf{U}_A$) estimated for a simulated system with four constituents, calibrated with the spectra of 20 samples having random concentrations in the range 0 to 1. The order of loadings is: 1, blue; 2, green; 3, red; 4, light-blue

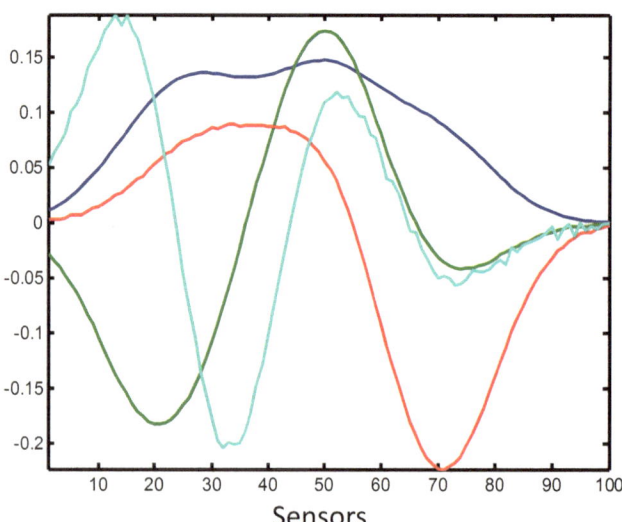

Fig. 7.2 First four PLS weight loadings (matrix $\mathbf{W}_A$) estimated for a simulated system with four constituents, calibrated with the spectra of 20 samples having random concentrations in the range 0 to 1. The order is as in Fig. 7.1

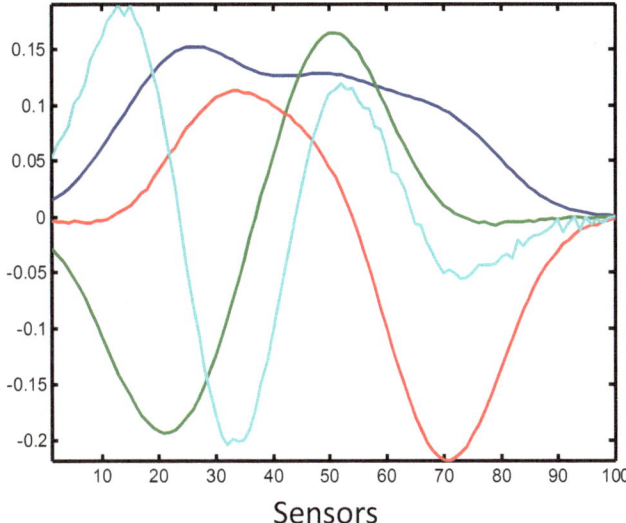

Fig. 7.3 First four PLS loadings (matrix $\mathbf{P}_A$) estimated for a simulated system with four constituents, calibrated with the spectra of 20 samples having random concentrations in the range 0 to 1. The order is as in Fig. 7.1

Table 7.1 Explained spectral variance by successive latent variables in a simulated system

	Explained spectral variance (%)	
Latent variable	PCR	PLS
1	91.3	91.2
2	6.8	6.8
3	1.5	1.6
4	0.4	0.4

close to each other, although not identical. The cumulative explained variance for the first 4 variables is in both cases ca. 100%, because this is an ideal system with four sample constituents and a low noise level.

As regards the vector of regression coefficients, Fig. 7.4 compares the PCR and PLS $\mathbf{b}_n$ vectors for the analyte of interest in the above system: they are virtually identical. Is this a suggestion that PCR and PLS do not significantly differ in predictive power? Maybe. However, this result corresponds to a single and rather simple analytical system, with a low level of noise, and the conclusion might not be general. Nevertheless, recent research tends to confirm the predictive equivalence of both models, as previously discussed.

After the model is built, validation is performed as already described for previous models.

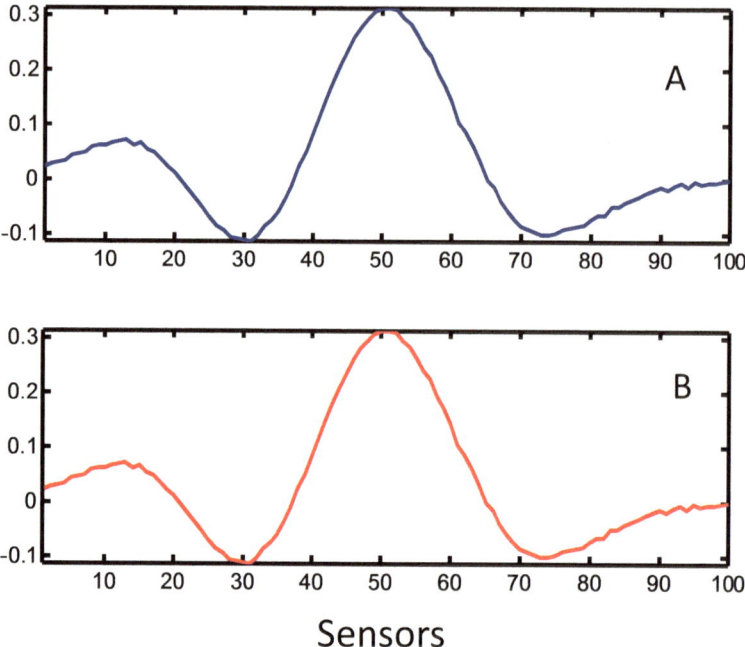

Fig. 7.4 Vectors of regression coefficients estimated by PCR (A) and PLS (B) estimated for a simulated system with four constituents, calibrated with the spectra of 20 samples having random concentrations in the range 0 to 1

7.6   A PLS Algorithm

Box 7.1 shows an iterative PLS algorithm. It estimates the loadings and scores one by one, in the order of their contribution to the spectral variance and of the covariance with the analyte concentration. In this sense, it reminds a PCR algorithm which, in the first PCA phase, estimates loadings and scores in the same fashion: one by one, in order of their contribution to the spectral variance. The latter is known as NIPALS (non-linear iterative partial least-squares) (Wold 1966).

Box 7.1
A PLS algorithm. The input variables for this algorithm are 'X', the matrix of calibration spectra, 'yn', the vector of calibration concentrations for the analyte or property of interest, 'A', the number of latent variables for building the model, and 'x', the spectrum for the unknown sample. The output variable is 'y', the predicted analyte concentration in the test sample.
for i=1:A

(continued)

Box 7.1 (continued)

```
        w=X*yn/(yn'*yn);
        W(:,i)=w/sqrt(w'*w);
        T(:,i)=X'*W(:,i);
        v(:,i)=T(:,i)'*yn/(T(:,i)'*T(:,i));
        P(:,i)=X*T(:,i)/(T(:,i)'*T(:,i));
        X=X-P(:,i)*T(:,i)';
        yn=yn-v(:,i)*T(:,i) ;
    end
    bn=W*inv(P'*W)*v';
    y=bn'*x;
```

The size of the variables are 'X', $J \times I$, (J = number of wavelengths or sensors, I = number of calibration samples), 'y', $I \times 1$, 'A', 1×1, and 'x', $J \times 1$. Those for the generated variables during program execution are: 'w', $J \times 1$, 'W', $J \times A$, 'T', $I \times A$, 'v', $A \times 1$, 'P', $J \times A$, 'bn', $J \times 1$ and 'y', 1×1.

We may compare Box 7.1 with the codes for ILS (Box 3.1 of Chap. 3) and PCR (Box 5.5 of Chap. 5). The first impression is that the complexity increases from ILS to PCR to PLS. It comes from simply *counting lines*. As usual, the first impression might be false. In ILS, the variable 'X' is not the original data matrix, but a considerably smaller matrix, obtained by selecting a suitable sub-set of wavelengths. This may lead to a separate variable selection MATLAB code, which may be much more complex than the mere ILS lines of Box 3.1. Hence, a combination of the simple ILS code and a variable selection algorithm may place ILS at the top of the complexity list.

On the other hand, PCR invokes the sub-routine 'princomp', which estimates the principal components, and makes the whole algorithm more complex than is suggested by the visual inspection of Box 5.1. It is remarkable that the PLS codes of Box 7.1 only include simple mathematical operations: sums, products, and divisions, except perhaps the matrix inversion in the second last line. However, this inversion is trivial, so that from the point of view of programming simplicity PLS appears to be the best of the three competitors.

7.7   The First-Order Advantage

PLS also provides a model for the spectrum of the unknown sample, allowing one to compute spectral residues, analogous to those in PCR. The sample spectrum is reconstructed from the calibration loadings and the sample scores as:

$$\mathbf{x}_A = \mathbf{P}_A \mathbf{t}_A \tag{7.8}$$

from which the standard deviation of the spectral residues can be obtained:

$$s_{res} = \sqrt{\frac{\sum\limits_{j=1}^{J} \left(x_j - x_{Aj}\right)^2}{J - A}} \tag{7.9}$$

As in all models achieving the first-order advantage, if the value of s_{res} for a given sample is significantly larger than the level of instrumental noise, the sample is suspected of containing unexpected constituents, not taken into account in the calibration phase. This may require re-calibration, by adding to the calibration set new samples, representative of the composition of the unknowns.

To adequately assess whether the spectral residues are significantly large, a statistical F-test has been proposed. The test compares the squared residuals for the test sample and the average squared residuals of the calibration set. Specifically, the experimental F_{exp} value is computed as (Thomas and Haaland 1988):

$$F_{exp} = \frac{I \sum\limits_{j=1}^{J} e_j^2}{\sum\limits_{j=1}^{J} \sum\limits_{i=1}^{I} e_{ij}^2} \tag{7.10}$$

where I is the number of calibration samples, e_j is the element (at wavelength j) of the residual vector $\mathbf{e} = \mathbf{x} - \mathbf{x}_A$ for the test sample in question, and e_{ij} are the corresponding spectral residues at wavelength j for the calibration sample i. The value of F_{exp} in Eq. (7.10) is then compared with the critical F_{crit} value with I and $I \times J$ degrees of freedom.

A nice illustration of the use of Eq. (7.10) is provided in a literature work (Fernández Pierna et al. 2015). NIR spectroscopy was employed at the entrance of a feed mill to provide early evidence of non-conformity and unusual ingredients. The study focused on the characterization of pure soybean meal, detecting and quantifying unusual ingredients in the soybean meal, such as melamine, cyanuric acid, or whey powder (milk serum). The use of Eq. (7.10) allowed the authors to flag the samples having the undesired constituents with high efficiency.

Incidentally, it is worth mentioning here that two similar F-tests can be implemented during the cross validation phase to flag calibration outliers (Thomas and Haaland 1988). One is based on spectral residuals, as Eq. (7.10), and the other one on concentration residuals for the left out samples (Thomas and Haaland 1988). These two tests will be discussed in a future chapter devoted to the selection of calibration samples. In any case, as has previously been indicated, outliers should be carefully treated, as they may not be really so, and may require additional studies and experimental actions before removing them from the calibration set.

7.8   Real Cases

We first comment in this section on a real case, already studied in the context of CLS: samples containing mixtures of three active principles (rifampicin, isoniazid, and pyrazinamide) were employed to calibrate multivariate models for their determination in pharmaceutical forms (Goicoechea and Olivieri 1999). The CLS model, as discussed in Chap. 2, was unsuccessful, mainly due to the high level of spectral overlapping among the constituent spectra (Fig. 7.5). Using a PLS model, on the other hand, led to satisfactory results (Goicoechea and Olivieri 1999). The analytical results for an independent validation set were considered satisfactory, with average relative errors of 1.9%, 9.2%, and 2.1% for rifampicin, isoniazid, and pyrazinamide, respectively. The largest value corresponded to isoniazid, in agreement with the high degree of overlapping among the spectrum for this particular analyte and the remaining sample constituents.

In commercial pharmaceutical forms, the recoveries of the three active principles were in the order of 108%, 96%, and 94% with respect to the content declared by the manufacturers, and within the limits accepted by pharmaceutical norms for quality control. This development could allow for the simultaneous determination of the active principles in a commercial form, considerably faster than a chromatographic procedure, and without using toxic organic solvents.

Of course, PLS can be employed to build much more complex calibration models, such as those previously discussed for NIR studies of foodstuff, fuels, and a large variety of industrially important materials. For example, in Sect. 5.11 of Chap. 5 we studied the determination of various quality parameters of corn seeds using PCR. We may now compare the results with those provided by PLS. Table 7.2 shows the results, including optimum number of latent variables, estimated by LOO cross validation up to 25 latent variables (the calibration set consisted of 50 samples), average prediction errors for a validation set (RMSEP), and relative error of

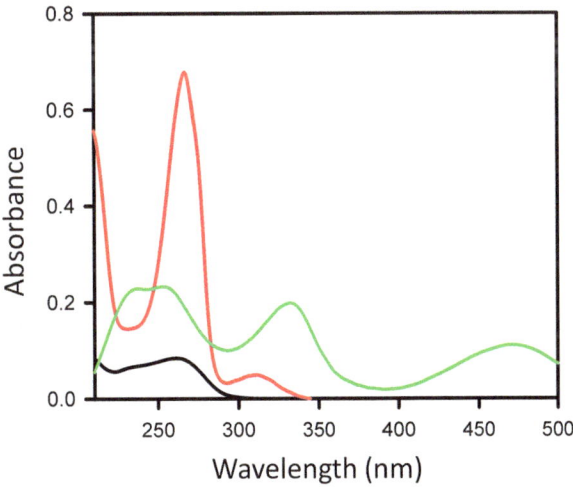

Fig. 7.5 UV-visible absorption spectra of the three pure constituents of a real sample: rifampicin (green line, 6.80×10^{-6} mol L^{-1}), isoniazid (black line, 2.00×10^{-5} mol L^{-1}), and pyrazinamide (red line, 1.20×10^{-4} mol L^{-1})

Table 7.2 Comparison of PCR and PLS validation results in the analysis of seeds[a]

	PCR			PLS		
Parameter	A	RMSEP/%	REP/%	A	RMSEP/%	REP/%
Moisture	23	0.016	0.16	18	0.014	0.14
Total oil	15	0.08	2.0	21	0.04	1.1
Total protein	25	0.11	1.2	18	0.11	1.2
Starch	22	0.17	0.3	21	0.19	0.3

prediction with respect to the mean calibration values (REP). They are as good as was already reported in Chap. 5, and for total oil they appear to be even better. Recall that the NIR spectra of these seed samples display a considerable amount of dispersion effects. The latter are better removed (or significantly decreased) before calibration with PCR and PLS, as will be explained in Chap. 10. Here the raw data were processed, and the results may not be representative of what is done in real practice with data of this type. Indeed, using the first derivative of the spectra instead of the raw spectra, PCR and PLS results for total oil are almost equivalent. The effect of spectral derivatives and other mathematical operations before building the models will be explained in Chap. 10. However, the data set is useful to show that for most of the studied parameters, PCR and PLS appear to be equivalent.

7.9   PLS-1 and PLS-2

Two different versions of PLS exist: PLS-1 and PLS-2. The algorithm described in Box 7.1. corresponds to PLS-1, and aims at calibrating a single analyte of interest. PLS-2, on the other hand, builds a model for several analytes at the same time. If an experimental system involves various analytes or sample properties, PLS-1 would model them separately, and would build a specific calibration model for each analyte or property. This is not problematic; on the contrary, the building of specific models, adapted to the needs of each analyte, is preferable to a single model for all analytes.

A PLS-1 model will require an optimum number of latent variables for each analyte, and will allow one to select, from the full spectra, a sub-set of spectral regions which might be appropriate for each analyte, as we shall see in a future chapter. In the same manner, if mathematical pre-processing is required before model building, PLS-1 will select a pre-processing which will also be analyte-specific.

The PLS-2 model is intended to calibrate all analytes with a single model, number of optimum latent variables, spectral region and data pre-processing. When is PLS-2 recommended? The answer can be found in the characteristics of PLS-2. In the calibration phase, PLS-1 correlates the data matrix $\mathbf{X}$ with the vector of analyte concentrations $\mathbf{y}_n$, whereas in PLS-2 the correlation is with the calibration matrix $\mathbf{Y}$ for all analytes. During the PLS-2 regression, $\mathbf{Y}$ is replaced by a matrix of Y-scores, just as $\mathbf{X}$ is replaced by the X-score matrix. This means that PLS-2 is preferable to PLS-1 when there are correlations in the columns of $\mathbf{Y}$, because the use of orthogonal Y-scores alleviates the correlation issues.

As an example, consider the calibration matrix $\mathbf{Y}$ containing the values of four quality parameters for corn seeds, studied in Chap. 5. Because the values of moisture, total oil, total protein, and starch have different scales, it is first convenient to transform each column to a new scale ranging from 0 to 1. Once scaled in this manner, the % of explained variance by the first four principal components of the matrix $\mathbf{Y}$ are estimated as 59%, 18%, 15%, and 8%. Because all four principal components are required to explain the variance, and the value for the fourth principal component is significant (8%), PLS-1 should be the model of choice.

On the other hand, if a matrix $\mathbf{Y}$ for four analytes, after scaling the columns, requires *three* principal components for explaining most of the variance, and the fourth principal component only explains a very small %, PLS-2 should be selected for model building. The modeling of $\mathbf{Y}$ by three principal components implies the presence of a significant degree of correlation, for which PLS-2 is better suited.

In sum, there is no point in using both PLS-1 and PLS-2 for a given system. The logical procedure would be to first perform a PCA of the matrix $\mathbf{Y}$, and verify whether the number of principal components required to explain most of the variance in this matrix is equal or smaller than the number of $\mathbf{Y}$ columns (the number of analytes). If it is equal, select PLS-1; if it is smaller, select PLS-2 (Wold et al. 2001).

In what follows, we will use the term PLS for PLS-1. When required, PLS-2 will be specifically named.

7.10   A PLS-2 Algorithm

A MATLAB algorithm for PLS-2 is given in Box 7.2. It looks a bit more complex than the one for PLS-1 in Box 7.1, due to the need of estimating both Y- and X-scores for calibration.

Box 7.2

A PLS-2 algorithm. The input variables are similar to Box 7.1, except that the matrix of analyte concentrations 'Ycal' replaces the single vector 'yn' in PLS-1, and the matrix of predicted test concentrations 'Y' replaces the vector 'y'.

```
for i=1:A;
u=Ycal(:,1);
w=X'*u/(u'*u);
Wn(:,i)=w/norm(w);
told=X*Wn(:,i);
T(:,i)=told;
```

(continued)

Box 7.2 (continued)

```
        v(:,i)=T(:,i)'*Ycal/(T(:,i)'*T(:,i));
        u=Ycal*v(:,i)/(v(:,i)'*v(:,i));
        w=X'*u/(u'*u);
        Wn(:,i)=w/norm(w);
        tnew=X*Wn(:,i);
        while norm(told-tnew)>1e-6*told
            w=X'*u/(u'*u);
            Wn(:,i)=w/norm(w);
            told=X*Wn(:,i);
            T(:,i)=told;
            v(:,i)=T(:,i)'*Ycal/(T(:,i)'*T(:,i));
            u=Ycal*v(:,i)/(v(:,i)'*v(:,i));
            w=X'*u/(u'*u);
            Wn(:,i)=w/norm(w);
            tnew=X*Wn(:,i);
        end
        P(:,i)=X'*T(:,i)/(T(:,i)'*T(:,i));
        X=X-T(:,i)*P(:,i)';
        Ycal=Ycal-T(:,i)*v(:,i)';
    end
    b=Wn*inv(P'*Wn)*v';
    Y=b*x;
```

7.11   PLS: Discriminant Analysis

In Chap. 4, we described the application of PCA to the discrimination of samples. We called the procedure *unsupervised*, implying that the samples automatically separated in groups, based on the latent variables estimated only from the data matrix $\mathbf{X}$, and without previous knowledge on the existence of groups or sample classes. PLS can also be used for similar purposes, but in this case the analysis is called *supervised*, because this model requires the knowledge of the vector $\mathbf{y}_n$ of target properties (in this case the classes to which the samples belong). The procedure is known as PLS-discriminant analysis (PLS-DA).

To build a PLS-DA model, one should have a set of signals (spectra or other multivariate signals) for a number of samples, and knowledge of the samples classes instead of analyte concentrations. Suppose we have measured the spectra of 20 different samples, distributed between two classes with 10 samples each, as was previously studied with PCA in Chap. 4. In this case we positively know that the first ten samples belong to class A, and the ten subsequent samples to class B. To build the PLS model between $\mathbf{X}$ and class A, we employ a vector $\mathbf{y}_n$ of size 20×1, whose first 10 elements are 1 (the class A code), and the next 10 elements are 0 (the code for the remaining class or classes):

$$
\mathbf{y}_n = \begin{bmatrix} 1 \\ 1 \\ 1 \\ \cdots \\ 1 \\ 0 \\ 0 \\ 0 \\ \cdots \\ 0 \end{bmatrix} \quad \begin{matrix} \text{Class}\ \ \text{A} \\ \text{Class}\ \ \text{A} \\ \text{Class}\ \ \text{A} \\ \cdots \\ \text{Class}\ \ \text{A} \\ \text{Class}\ \ \text{B} \\ \text{Class}\ \ \text{B} \\ \text{Class}\ \ \text{B} \\ \cdots \\ \text{Class}\ \ \text{B} \end{matrix} \tag{7.11}
$$

The first phase of PLS model building is LOO cross validation, to establish the optimum number of latent variables, which in this case is two. The predicted values for each sample left out during cross validation may be useful to judge the success of the calibration model. These values are shown in the bar plot of Fig. 7.6. As can be seen, the predicted values for class A are all above the limit separating the classes (0.5 units), while the remaining values are all well below the limit (class B). This

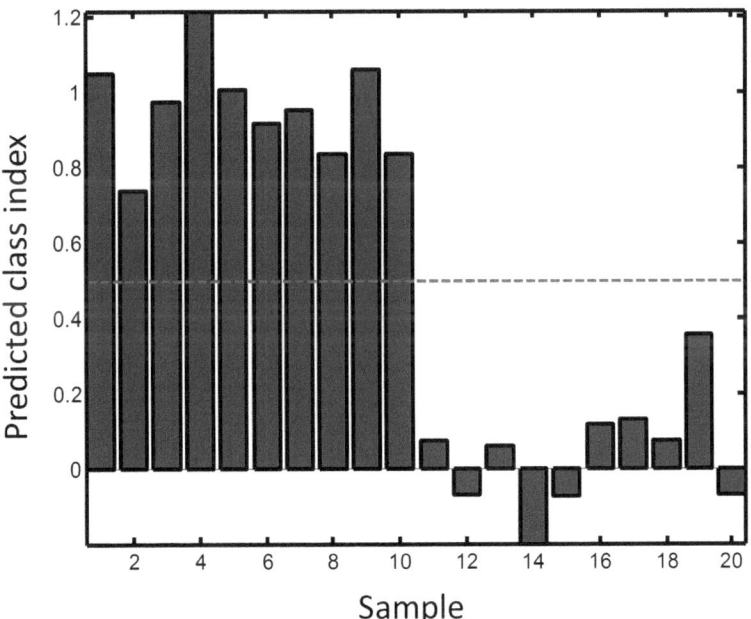

Fig. 7.6 Bar plot showing the predictions by a PLS-DA model during cross validation of a set of 20 calibration samples (see Chap. 4, Sect. 4.6). The predicted values correspond to the class index for each sample, originally designated as 1 (class A) and 0 (class B). The red dashed line indicates the separation limit between classes (0.5 units)

supervised calibration can be considered successful, at least concerning the cross validation phase. One should then analyze an independent validation set of samples. Once the PLS regression coefficients for the calibration model are found, they can be used to predict the class index for future samples. If this value is close to 1, the sample will be classified as A, if it is close to 0, to class B. This rule allows one to call the entire process a *classification*.

7.12   Another Real Case

In this classification application, a handheld NIR device was used for classifying six similar Amazonian woods (Soares et al. 2017). Supervising wood exploitation can be very challenging due to the existence of many similar species and the reduced number of wood identification experts to meet the demand. There is evidence that valuable endangered wood species are being smuggled disguised as other species. The studied species were mahogany (*Swietenia macrophylla*), cedar (*Cedrela odorata*), both of high value, crabwood (*Carapa guianensis*), cedrinho (*Erisma uncinatum*), curupixá (*Micropholis melinoniana*), and jatobá (*Hymenea coubaril*).

The NIR spectra of 113 samples of wood were measured each 8 cm^{-1} in the range between 9000 and 4000 cm^{-1}. The calibration was made with 68 samples, leading to four different PLS models (one for separating each class from the remaining ones), and validation was carried out on the remaining 45 samples. Figure 7.7 shows the average NIR calibration spectra for each species, and Fig. 7.8 the calibration and

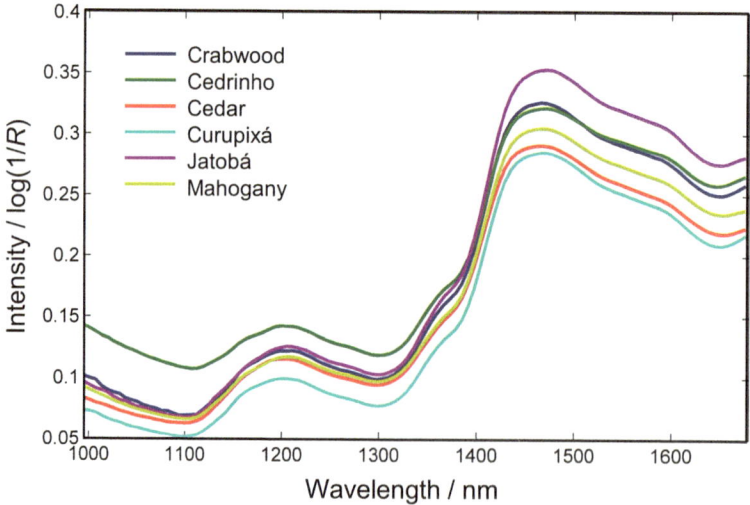

Fig. 7.7   Average NIR calibration spectra for six different types of raw woods, employed to build a PLS-DA calibration model to classify wood samples. The vertical scale is the logarithm of the inverse of the reflectance values. Adapted with permission from Soares et al. (2017) (Brazilian Chemical Society)

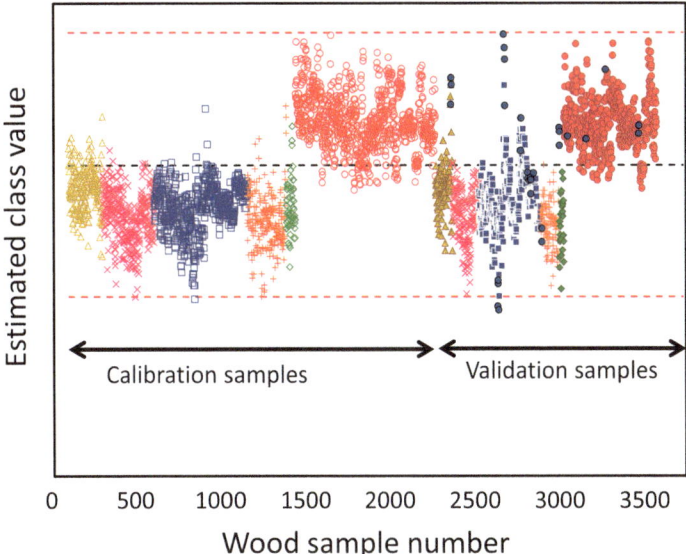

Fig. 7.8 Estimated class index values in the calibration phase (open red circles) and in the validation phase (filled red circles) for the classification of the high-value mahogany samples from the remaining ones (all other symbols). Black circles indicate a few anomalous samples. The dotted lines are the limits for class identification (Soares et al. 2017). Adapted with permission from Soares et al. (2017) (Brazilian Chemical Society)

validation results for classifying mahogany. Similar results were obtained for distinguishing each species from the remaining five other classes. From Fig. 7.8, one may appreciate the excellent discriminant capacity of PLS-DA, allowing one to predict the wood classes by a non-invasive spectroscopic method in real time. It should be noticed that the classical methodology requires expert personnel with training in wood taxonomy, which is difficult in field studies, where immediate response is required. The NIR/PLS-DA methodology provides the correct answer, by means of a portable and inexpensive NIR spectrometer.

7.13   Advantages of PLS

The advantages of PLS are the same as those for PCR, with the additional benefit that the latent variables are, in principle, analyte-specific, and thus better adapted to the needs of each analyte of interest. Some years ago, the author of this book wrote a scientific work entitled *Ten reasons to prefer PLS to ILS*. It was not accepted, and I admit to have exaggerated in one or two items. However, my position is still firm; the ten reasons are the following ones.

1. More reliable analytical predictions and calibration models
2. Extremely useful diagnostic information (the first-order advantage)

3. Relative decrease of the impact of noise
4. Higher sensitivity
5. Lower detection and quantitation limits (to be discussed in a future chapter)
6. Algorithmic simplicity
7. Full-spectral regression coefficients with useful qualitative information
8. Zero-correlation scores
9. Valuable insight into the mutual relationship among samples (to be discussed in a future chapter)
10. Visual (and statistical) inspection of the signal–concentration relationship
11. As long as these arguments do also apply to PCR, the ten reasons can also be extended to PCR

7.14 Beyond PLS

PLS is most probably the de facto first-order multivariate calibration standard. What can possibly go wrong with PLS? In the last years several competitors of PLS have been developed, from simple cosmetic variants to completely different methodologies. One idea which underlies some alternative PLS models is to first pre-process the original data matrix, with the aim of removing information which is not relevant regarding the analyte concentration or sample property to be measured. In this way, PLS receives a filtered data matrix, richer in information on the target analyte or property. These alternative models usually require fewer latent variables than the original one, because some variance sources have been previously removed. Although they are more parsimonious and perhaps preferable in this latter sense, the overall prediction results do not appear to be significantly better than traditional PLS. The reader may find additional details on these models in the specific literature (Svensson et al. 2002; Wold et al. 1998; Fearn 2000; Goicoechea and Olivieri 2001; Xu and Schechter 1997).

Another important assumption of all models so far is the linearity in the signal–concentration relationship. PLS will work properly if the relationship is linear, or if it does not appreciably deviate from the strict linearity. When this condition is not met, and significant non-linearity in the signal–concentration relation occurs, the PLS model will lose predictive power. In general, and due to the existence of Lambert–Beer's law, in analytical spectroscopy the relationship is fairly linear in a certain concentration range. However, when measuring other multivariate signals, or when calibrating for global sample properties (octane number in fuels, acceptability of a product by a sensor panel, etc.), the relation may not be linear. In cases of significant non-linearity, alternative models are needed. There are many different models of this type; a popular group includes the so-called artificial neural networks (ANN) (Ni et al. 2014). We will describe in a future chapter a specific ANN variant, discussing whether it is justified to employ non-linear models or it is preferable to maintain the simpler PLS model.

Finally, some progress is related to the characteristics of the instrumental noise. Until now we have implicitly considered that the instrumental noise is not only

Fig. 7.9 Three different types of instrumental noise: (**a**) iid, (**b**) correlated, and (**c**) proportional

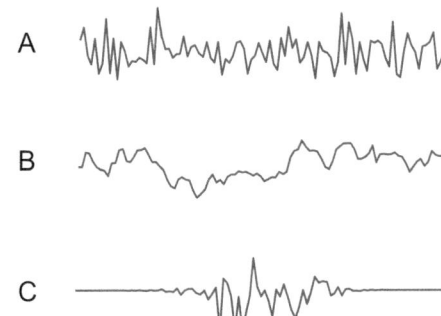

random, but also independent and identically distributed (iid). This implies that there are no correlations among the noise at one wavelength and the noise at the neighboring wavelengths (independent noise), and that the standard deviation of the noise is constant across the spectrum (identically distributed noise). Figure 7.9 shows three traces of noisy signals, illustrating the following noise types: iid (A), correlated or not independent (B), and proportional or not constant (C).

When we apply the criterion of least-squares in all the regression problems, the noise is assumed to be iid, so that all models considered so far are based on the same assumption. However, the noise not always behaves in this way (Wentzell 2014). For non-iid noise, the models based on the iid assumption will be sub-optimal, and alternative ones taking into account the noise structure should show improved predictive ability. One of such models has been developed based on PCR, but following a criterion called maximum likelihood (MLPCR) (Schreyer et al. 2002). The latter has not gained a significant popularity, due to the difficulties associated with the estimation of a key parameter describing the noise structure: the so-called error variance-covariance matrix (Schreyer et al. 2002). In any case, Reis and Saraiva (2004) have shown that PLS is able to resist the presence of non-iid noise, except perhaps when significant correlations among the noise in neighboring sensors exist.

7.15 Exercises

1. (a) Show the mathematical steps needed to obtain Eq. (7.2) from Eq. (7.1)
   (b) Find the specific definition for the generalized inverse $\mathbf{T}_A^+$ in Eq. (7.2)
2. Discuss the mathematical requirements for solving Eq. (7.1) by least-squares, in terms of the number on independent equations and unknowns
3. A discriminant PLS model is built for separating 5 samples of class A, 4 samples of class B, and 6 samples of class C. Write the three $\mathbf{y}_n$ vectors required for separating each class from the remaining ones, if the data matrix was generated with the corresponding spectra in the following way: the first 5 columns of $\mathbf{X}$ for class A samples, the next 4 columns for class B, and the final 6 columns for class C

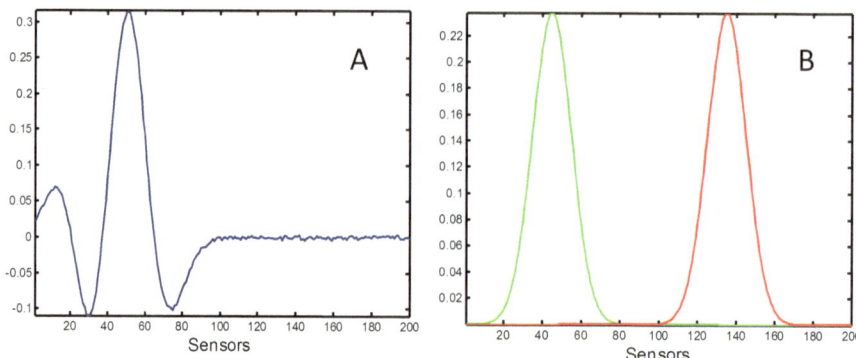

Fig. 7.10 (**a**) Spectrum of a vector of PLS regression coefficients for a typical analytical system. (**b**) Spectra of potential interferents, not considered in the calibration phase, that may appear in future unknown samples

Table 7.3 RMSEP values for different multivariate models

Model	Analyte 1	Analyte 2	Analyte 3
ILS	0.189	0.151	0.465
PLS-1	0.014	0.013	0.016
PLS-2	0.015	0.014	0.017

4. Three active principles of a pharmaceutical formulation need to be determined with a suitable PLS model from spectral data. Justify the expected optimum value of A if the spectral data correspond to:
   (a) UV-visible absorption spectroscopy in solution
   (b) NIR spectroscopy in the solid tablets
5. The vector of regression coefficients for a PLS model of an analytical system is shown in Fig. 7.10 A. When studying new samples, unexpected interferents may appear, whose spectra are shown in Fig. 7.10 B
   (a) Will the samples containing the interferent with the red trace be flagged as an outlier according to Eq. (7.10)?
   (b) Will the analyte concentration be predicted with a significant bias in the presence of the red interferent?
   (c) Answer the above two questions for the interferent with the green trace in Fig. 7.10 B
6. How problematic would you expect to be a sample with constituent concentrations which are far from the center of the calibration range, but with no new constituents?
7. Table 7.3 collects the RMSEP results for three analytes using the linear models ILS, PLS-1, and PLS-2.
   (a) Are the numbers reported with appropriate significant figures? How would you report them?
   (b) What can you conclude at first sight on the application of the three models?

References

Fearn, T.: On orthogonal signal correction. Chemom. Intell. Lab. Syst. **50**, 47–52 (2000)

Fernández Pierna, J.A., Abbas, O., Lecler, B., Hogrel, P., Dardenne, P., Baeten, V.: NIR fingerprint screening for early control of non-conformity at feed mills. Food Chem. **189**, 2–12 (2015)

Goicoechea, H.C., Olivieri, A.C.: Simultaneous determination of rifampicin, isoniazid and pyrazinamide in tablet preparations by multivariate spectrophotometric calibration. J. Pharm. Biomed. Anal. **20**, 681–686 (1999)

Goicoechea, H.C., Olivieri, A.C.: A comparison of orthogonal signal correction and net analyte preprocessing methods. Theoretical and experimental study. Chemom. Intell. Lab. Syst. **56**, 73–81 (2001)

Ni, W., Nørgaard, L., Mørupc, M.: Non-linear calibration models for near infrared spectroscopy. Anal. Chim. Acta. **813**, 1–14 (2014)

Reis, M.S., Saraiva, P.M.: A comparative study of linear regression methods in noisy environments. J. Chemom. **18**, 526–536 (2004)

Schreyer, S.K., Bidinosti, M., Wentzell, P.D.: Application of maximum likelihood principal components regression to fluorescence emission spectra. Appl. Spectrosc. **56**, 789–796 (2002)

Soares, L.F., da Silva, D.C., Bergoa, M.C.J., Coradina, V.T.R., Braga, J.W.B., Pastore, T.C.M.: Avaliação de espectrômetro NIR portátil e PLS-DA para a discriminação de seis espécies similares de madeiras amazônicas. Quim Nova. **40**, 418–426 (2017)

Svensson, O., Kourti, T., MacGregor, J.F.: An investigation of orthogonal signal correction algorithms and their characteristics. J. Chemom. **16**, 176–188 (2002)

Thomas, E.V., Haaland, D.M.: Partial least-squares methods for spectral analyses. 1. Relation to other quantitative calibration methods and the extraction of qualitative information. Anal. Chem. **60**, 1193–1202 (1988)

Wentzell, P.D.: Measurement errors in multivariate chemical data. J. Braz. Chem. Soc. **25**, 183–196 (2014)

Wold, H.: Estimation of principal components and related models by iterative least squares. In: Krishnaiah, P.R. (ed.) Multivariate Analysis, pp. 391–420. Academic Press, New York (1966)

Wold, S., Antti, H., Lindgren, F., Öhman, J.: Orthogonal signal correction of near-infrared spectra. Chemom. Intell. Lab. Syst. **44**, 175–185 (1998)

Wold, S., Sjöström, M., Eriksson, L.: PLS-regression: a basic tool of chemometrics. Chemom. Intell. Lab. Syst. **58**, 109–130 (2001)

Xu, L., Schechter, I.: A calibration method free of optimum factor number selection for automated multivariate analysis. Experimental and theoretical study. Anal. Chem. **69**, 3722–3730 (1997)

Sample and Sensor Selection

8

Abstract

Multivariate calibration models are usually implemented by first selecting appropriate calibration samples and working wavelengths. Different procedures are discussed for performing these important activities.

8.1 Pre-calibration Activities

In the simplest situation, the analyst collects samples for calibration and validation, builds the multivariate model with the calibration data using the optimum number of latent variables, and checks the prediction performance against the validation samples. The model is then applied for analyte prediction in future unknown samples.

In general, however, there are various activities which are usually conducted prior to model building. They concern: (1) the specific samples to be used for the calibration and validation sets, (2) the spectral wavelengths (variables or sensors) to be used for modeling, and (3) whether raw or mathematically pre-processed data will be submitted to model building. The components of this triad, samples, sensors, and pre-processing, will be separately discussed in this and the next chapter, but they are mutually connected, so that one really does not know what comes first, *if the chicken or the egg*. This chapter is devoted to sample and sensor selection; we discuss mathematical pre-processing in the next chapter.

8.2 Sample Selection

When the chemical composition of the calibration samples is dictated by an adequate statistical design, the properties of the design themselves guarantee the sample representativity which is always required for model building. On the other hand,

© Springer Nature Switzerland AG 2018

A. C. Olivieri, *Introduction to Multivariate Calibration*,

https://doi.org/10.1007/978-3-319-97097-4_8

Fig. 8.1 Graphical illustration of a set of calibration samples and its distribution in the sample space. The red circles indicate samples which are far from the calibration center

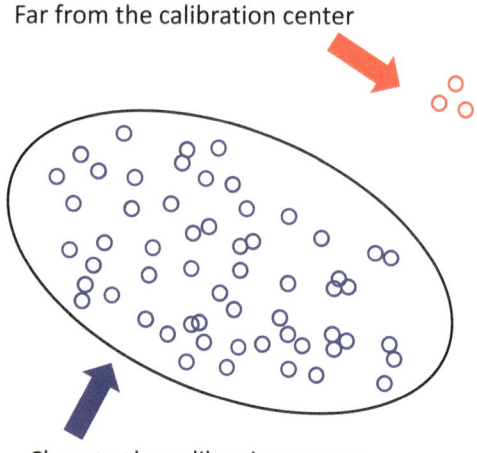

Far from the calibration center

Close to the calibration center

the validation sample set is usually selected at random, i.e., with constituent concentrations which are random numbers, constrained to be within the range of applicability of the model.

Nevertheless, an important number of multivariate calibration applications cannot be developed by designing the training sample set with specific properties under the control of the analyst. The samples may come from an industrial production line, or they may be of natural origin, and cannot be controlled as regards composition or physical properties. There are multiple examples of this case, involving the determination of quality parameters of food stuff, drinks, cosmetic, textile or paper products, fuels, etc.

When the sample composition cannot be controlled, and a large number of samples is available, the following problem arises: how many and which samples should be employed for calibrating the model, and how many and which should be left for the validation phase. A usual procedure is to randomly divide the whole sample set, leaving, for example, 70% for calibration and 30% for validation. Is there any risk in this random selection of samples? In general, yes, because random selection may leave out samples which are far from the calibration center, or correspond to regions of the sample space which are relatively less populated. Figure 8.1 illustrates this fact: if the red samples are not chosen by a random selection method, the calibration set will only be represented by the blue specimens, and will not display the necessary representativity regarding future red samples.

To avoid these problems, sample selection algorithms have been developed, which keep in the calibration set a number of really representative objects. One of the most popular ones is the Kennard–Stone algorithm (Kennard and Stone 1969), which can be summarized in the following steps.

1. The first selected sample is the one closest to the center of the sample space.
2. The second selected sample is the most distant to the first one.

3. For the subsequent selections, the distances of the remaining samples to those previously selected are first computed. Then the minimum distances are considered, and the chosen sample is the one with the largest of the minimum distances.
4. The process continues until a certain pre-defined number of calibration samples, e.g., 70% of the total number. The remaining ones are left for validation.

In the above steps, the distance $d_{ii'}$ between samples i and i' is defined by:

$$d_{ii'} = \sqrt{\sum_{j=1}^{J} (x_{ji} - x_{ji'})^2} \tag{8.1}$$

where x_{ji} and $x_{ji'}$ are the values of the signal for samples i and i' at wavelength j. Eq. (8.1) should not come as a surprise: it is the generalization of the Pythagorean theorem to a multi-dimensional space, giving the *hypotenuse* of a generalized hyper-triangle in J dimensions. The Kennard–Stone method first selects a central object, then objects near the border of the sample space, and then fills the space with the remaining objects in an orderly way.

As an example, consider the 10 samples represented by blue circles in Fig. 8.2, whose positions in the sample space are defined by only two coordinates [$J = 2$ in Eq. (8.1)]. The first selected sample is indicated in Fig. 8.2, and corresponds to the one closest to the center (the red cross marks the center). The second one will be that located further away from the first one, as also indicated in Fig. 8.2.

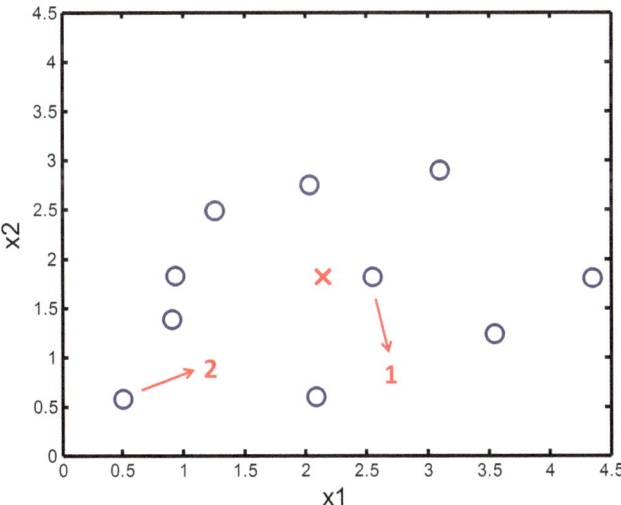

Fig. 8.2 The Kennard–Stone algorithm in action: from 10 samples distributed in a two-dimensional sample space, the first to be selected is the one closest to the calibration center (marked with a red cross), and the second is the one located farther away from the first

The distances of the remaining eight samples to the first two are then computed (two distances for each of the eight remaining samples). Figure 8.3 shows the minimum of these two distances per remaining sample (green lines). The solid green line in this figure is the maximum of the eight minimum distances, and defines the third selected sample, as indicated. The process continues in the same fashion. Notice that the second selected sample, located at the largest distance from the center, is included for calibration by this algorithm, but might not be selected by a random selection method.

An illustration of the overall process is provided by Fig. 8.4, showing the 25 selected samples by the Kennard–Stone algorithm, from a set of 100 randomly

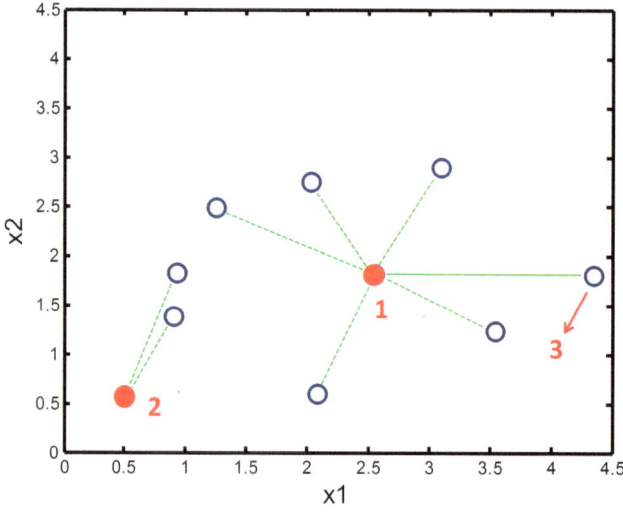

Fig. 8.3 The Kennard–Stone algorithm in action: once the first two samples are selected (red circles), the distance of each of the remaining samples to the first two are calculated, and each of the minimum of these distances is chosen, as indicated by green lines (both solid and dotted). The third selected sample is the one with the largest of the minimum distances (solid green line)

Fig. 8.4 Selection of 25 samples (red crosses) using the Kennard–Stone algorithm, from a set of 100 samples (blue circles), randomly distributed in a two-coordinate sample space

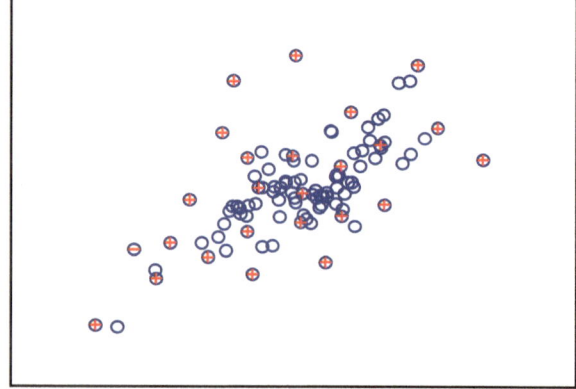

distributed samples. As can be seen, those in the external borders of the calibration space are consistently selected, followed by additional ones located inside the calibration area.

There are additional procedures for sample selection, including variants of the Kennard–Stone algorithm, which introduce concentration information in the selection process and not only signal information (Galvao et al. 2005). Many other methods, using seemingly different philosophies, have been developed, although Kennard–Stone is still a classic and reliable technique.

8.3   An Algorithm for Kennard–Stone Sample Selection

The Kennard–Stone algorithm can be implemented in MATLAB with the codes given in Box 8.1. The algorithm provides a vector containing the indexes of the selected samples and the corresponding matrix of signals for the latter samples.

Box 8.1

The Kennard–Stone algorithm. Input variables are 'X', the complete data matrix, and the value of 'Isel' (the number of samples to be selected as representative). The algorithm provides the vector 'msel' containing the indexes of the selected samples, and the matrix of signals for the selected samples 'Xsel'.

```
[min1,s1]=min(sum((X-mean(X,2)*ones(1,size(X,2))).^2));
[max1,s2]=max(sum((X-X(:,s1)*ones(1,size(X,2))).^2));
Xsel=[X(:,s1),X(:,s2)];
msel=[s1;s2];
for isel=1:Isel-2
    d=zeros(size(X,2),size(Xsel,2));
    for i=1:size(X,2)
        d(i,:)=sum((X(:,i)*ones(1,size(Xsel,2))-Xsel).^2);
    end
    [max2,sisel]=max(min(d'));
    msel=[msel;sisel];
    Xsel=[Xsel,X(:,sisel)];
end
disp(msel)
```

8.4   Calibration Outliers

Once the samples for the calibration set are selected, and their spectra or multivariate signals are measured, cross validation is applied to estimate the optimum number of latent variables. During this procedure, it is possible to detect the presence of outlying samples: those whose nominal analyte concentration or property significantly deviates from the prediction when they are left out from the set, relative to the average cross validation error.

An indicator for the presence of outliers in cross validation has been proposed in the form of an $F_y(i)$ value for the ith calibration sample, defined by (Thomas and Haaland 1988):

$$F_y(i) = \frac{(I-1)\left(y_{\mathrm{pred},i} - y_{\mathrm{nom},i}\right)^2}{\sum_{i' \neq i} \left(y_{\mathrm{pred},i'} - y_{\mathrm{nom},i'}\right)^2} \tag{8.2}$$

where $y_{\mathrm{pred},i}$ and $y_{\mathrm{nom},i}$ are the predicted and nominal value for the left out sample during cross validation, and $y_{\mathrm{pred},i'}$ and $y_{\mathrm{nom},i'}$ are the corresponding values for the remaining samples. The degrees of freedom for studying the significance of $F_y(i)$ are 1 and $(I-1)$ for the numerator and denominator, respectively. As can be seen in Eq. (8.2), $F_y(i)$ compares the variance for each left out sample with respect to the average variance for the remaining samples used to build the cross validation models.

Equation (8.2) defines the concentration or y-outliers, to distinguish them from the x-outliers, samples whose location in spectral or score space is far from the remaining ones. To detect x-outliers in cross validation, a similar $F_x(i)$ indicator has been proposed:

$$F_x(i) = \frac{(I-1)\sum_{j=1}^{J} (x_i - x_{A,i})^2}{\sum_{i' \neq i} \sum_{j=1}^{J} \left(x_{i'} - x_{A,i'}\right)^2} \tag{8.3}$$

where x_i and $x_{A,i}$ are elements of the original spectrum and of the one reconstructed with A latent variables, corresponding to the left out sample, while $x_{i'}$ and $x_{A,i'}$ have the same meaning, but apply to the remaining samples used to build the cross validation models. The number of degrees of freedom for the significance of $F_x(i)$ is a matter of a certain controversy. A useful rule of thumb says that for a large number of wavelengths ($J \gg I$), values of $F_x(i)$ smaller than ca. 3 do not indicate the presence of significant outliers (Thomas and Haaland 1988).

Typical plots for outlier detection in calibration are shown in Fig. 8.5, displaying the values of the ratios F_y/F_{crit}, which allow one to conclude whether a given sample is suspicious of being an outlier. In the case of Fig. 8.5, a PLS model was built for the determination of octane number in gasolines from the NIR spectra of 48 samples

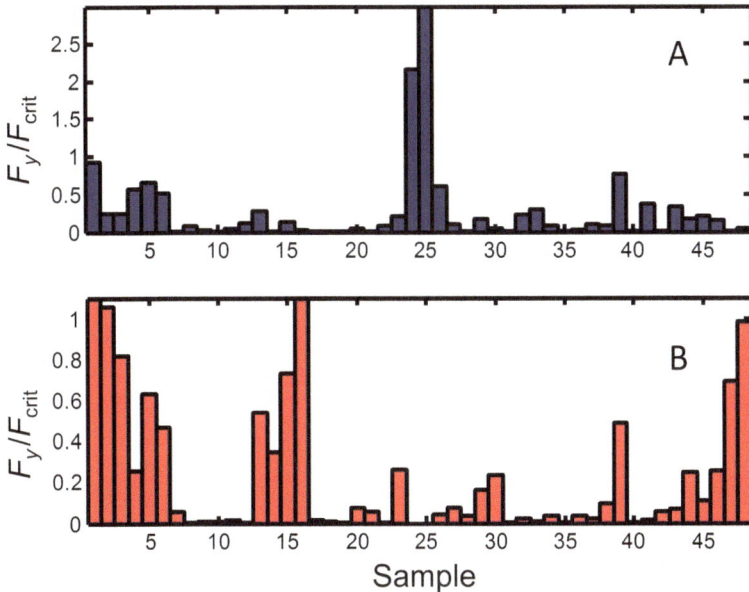

Fig. 8.5 Bar plots for the detection of y-outliers. (**a**) Values of the ratio F_y/F_{crit} for each calibration sample after LOO cross validation, indicating that samples 24 and 25 are outliers. (**b**) Values of the ratio F_y/F_{crit} after LOO cross validation, having removed samples 24 and 25 from the calibration set

with known octane number. Cross validation showed that the optimum number of latent variables was 5, with an RMSECV value of 0.58 octane units, but the plot for detection of y-outliers clearly suggests that samples 24 and 25 significantly deviate from the ideal predictions (Fig. 8.5a). Once these samples are removed from the calibration set, no new outliers are found, and a new cross validation process led to an optimum number of latent variables of 3, with an improved RMSECV value of 0.31 units.

Two important details should be taken into account when searching for outliers. First, one should not be too strict regarding the value of the ratio F_y/F_{crit}: if values larger than 1 (e.g., 1.2) occur, it is likely that removing these samples would not produce a significant effect in the quality of the calibration, especially if the calibration set contains a large number of samples. Second, if an outlier is real, removing it from the calibration should lead to a smaller number of optimum latent variables, accompanied by a decrease in the RMSECV, as in the example of Fig. 8.5.

As a colorful anecdote on the subject, in 2006 the author of this book attended an international meeting (X CAC, Chemometrics in Analytical Chemistry, Aguas de Lindoia, Brazil), and was in the audience when a researcher was talking about outliers. At the end of the conference, Prof. Josef Havel, from Masaryk University, Brno, Check Republic, raised his hand and said out loud: *there are no outliers*! What did Prof. Havel mean? I guess these two important concepts: (1) signal outliers should not be removed from calibration, but more samples of the same type should be included in the calibration to increase the representativity, and (2) concentration outliers should be prepared or analyzed again with the reference technique. In other words: outliers do not exist.

8.5   Sensor Selection

In previous chapters, we have highlighted the benefits of working with full-spectral inverse models, instead of with a few wavelengths. However, there is general consensus in that the predictive ability of multivariate models improves upon selecting a reduced range of spectral regions from the full spectra. This is not contradictory. The selection of sensors intends to build models based on a large number of sensors, much larger than for ILS models, but removing spectral regions which are not sensitive regarding the analyte or property of interest. A compromise is thus achieved of *many sensors but not too many*, in such a way that the model receives spectral regions which display significant sensitivity towards the analyte, discarding those where the analyte does not respond. Here we appreciate the value of the specific PLS-1 model, where a given spectral region corresponds to each analyte. In PLS-2, on the other hand, the same spectral regions should be employed for studying all analytes.

Many different methods exist for selecting sensors, all with the general objective of improving the predictive power, and new methods are proposed almost every day (Mehmood et al. 2012). In this chapter, we will explore in detail two of them: the selection based on the vector of regression coefficients, and the method called interval-PCR/PLS.

8.6   Regression Coefficients for Selecting Sensors

The basis for the selection of working wavelengths using the vector of regression coefficients can be found in the expression for predicting the analyte concentration y_n:

$$y_n = \mathbf{b}_n^{\mathrm{T}}\mathbf{x} = b_{1n}x_1 + b_{2n}x_2 + \ldots + b_{Jn}x_J \qquad (8.4)$$

If the elements of the vector $\mathbf{b}_n$ are in a certain spectral region close to zero and/or have a large uncertainty, the terms in Eq. (8.4) corresponding to these elements will be very small and/or present a significant uncertainty relative to the remaining terms. Therefore, a method has been proposed for wavelength selection which considers the spectrum of regression coefficients, estimated with the optimum number of latent variables, discarding the regions where the elements of $\mathbf{b}_n$ are negligible. Since the expression *negligible* needs statistical support, several procedures have been designed to choose spectral regions where the elements of $\mathbf{b}_n$ are statistically significant (Centner et al. 1996).

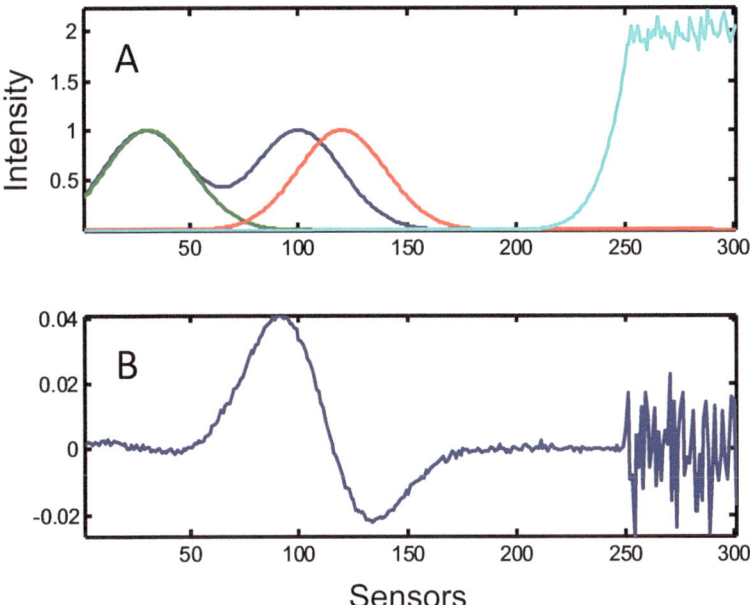

Fig. 8.6 (**a**) Ideal analytical system with an analyte of interest (blue spectrum), two additional constituents (red and green spectra), and a baseline saturating the detector in the sensor range from 250 and 300 (light-blue spectrum). (**b**) Vector of PLS regression coefficients for the analyte of interest in this system

Usually, however, the visual inspection of the spectrum of $\mathbf{b}_n$ is useful. Figure 8.6 shows a typical situation. Figure 8.6a shows the spectra of various constituents of an analytical system, including a baseline signal which saturates the detector. The calibration set was composed of 10 samples with random concentrations of three constituents in the range from 0 to 1 unit, including the baseline in all samples. The vector $\mathbf{b}_n$ is shown in Fig. 8.6b. What conclusions can be drawn from this example?

1. In the region where the detector is saturated (250–300), there is considerable effect of the noise over $\mathbf{b}_n$. This region will propagate a significant uncertainty to the estimation of the concentration by Eq. (8.4). Removing this type of regions is recommended.
2. The region where the signal is very small (180–250) leads to values of $\mathbf{b}_n$ which are also very small. They will not significantly contribute to Eq. (8.4). The impact of the signal noise will decrease in the model compression phase, so that discarding or maintaining this type of regions will cause almost irrelevant effects on prediction.
3. The region where the analyte significantly responds, even when its spectrum is partially overlapped with those of other constituents (50–180), will be useful to calibrate, yielding values of $\mathbf{b}_n$ with spectral aspect. These regions should be kept for model building.

4. The region where the analyte significantly responds, but its spectrum is almost
   completely overlapped with those for other constituents (1–50), will be almost
   useless for calibration, giving rise to values of $\mathbf{b}_n$ with low analytical value.
   Discarding or maintaining this type of regions will cause almost irrelevant effects
   on prediction.

We should notice that the above analysis was made on an ideal case, where all
constituent properties are known. What would happen in real life? We would not
know where the analyte responds, moreover, there might be no such thing as an
analyte in the classical sense. The good news is that the consideration of the vector of
regression coefficients is independent on our knowledge about the system
constituents: it tells us that there are regions of $\mathbf{b}_n$ which are uncertain and should
be discarded, others with spectral aspect which should be kept, and still others with
very low intensity which can be discarded or kept without serious consequences.

Finally, it is important to note that the use of regression coefficients for sensor
selection, for detecting the position of spectral regions which are sensitive to the
analyte, or for concluding with regard to the properties of the analyte, has been
criticized (Brown and Green 2009). The ultimate word has not been said in this
subject of apparent complexity.

8.7   Interval-PCR/PLS

Instead of considering a relatively abstract object such as the vector of regression
coefficients, alternative methods have been developed, directly pointing to the heart
of analytical chemistry: the lowest prediction error. One alternative is the comprehensive search of useful spectral sensors, but this may be prohibitive in terms of
time. A hint regarding this issue was provided in Sect. 3.8 of Chap. 3, when
discussing the selection of appropriate working wavelengths for the ILS model.

In the interval-PCR/PLS (i-PCR/PLS) method, the full spectral range is divided in
a certain number of sub-regions or intervals with a pre-defined width (Nørgaard et al.
2000). In each of these intervals, a separate model is built, with the corresponding
optimum number of latent variables, and then the average cross validation error
(RMSECV) is considered within each interval. Those sub-regions with the lowest
RMSECV values will be recommended for building the final model. Notice that it is
possible to couple non-consecutive intervals, if they lead to similarly low RMSECV
values.

A typical i-PLS plot is shown in Fig. 8.7, corresponding to the system described
in the previous section, now analyzed on the basis of the PLS regression vector. The
full spectral range comprises 301 sensors, and was divided into 20 intervals of
15 sensors each (discarding the last sensor). For applying LOO cross validation in
each interval, the interval sensors (15) must be larger than both the number of
calibration samples (10) and the maximum number of latent variables ($A_{\max}$). As
can be seen in Fig. 8.7, the region where the RMSECV is minimum includes the
spectral region where the vector of regression coefficients has a spectral-type shape

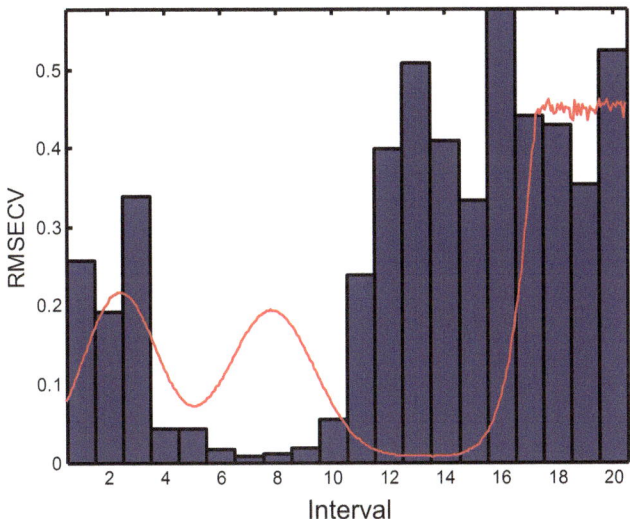

Fig. 8.7 Bar plot showing the cross validation error (RMSECV) for the system of Fig. 8.6 using the i-PLS method of variable selection with 15-sensor intervals. The red line shows the average calibration spectrum

(Fig. 8.6), i.e., the region between sensors 50 and 150, corresponding to the intervals numbered from 4 to 10. This result nicely agrees with the conclusions drawn from the behavior of the vector of regression coefficients in Sect. 8.6.

8.8 A Real Case

Figure 8.8 shows the visible-NIR spectra of a set of 61 sugarcane juice samples, all with known values of the degrees Brix, which were measured by means of the reference refractometric technique. Sugarcane juice is the syrup extracted from pressed sugarcane. The degrees Brix measure the relative content of dry matter (generally sugars) which is dissolved in the juice. As an example, a solution with 20 °Bx contains 20 g of dissolved solids per 100 g of solution. The Brix scale is used in the sugar industry to measure the approximate amount of sugars in the cane. The name comes from its proponent, Adolf Brix, a German engineer and mathematician (1798–1870).

The spectra of Fig. 8.8 show a region where the detector signal is saturated at ca. 1900 nm (reader: can you explain the presence of this intense NIR band?). It can be anticipated that a PLS model would find difficulties in building an accurate model if the latter band is included. In fact, LOO cross validation in the full spectral range leads to 12 as the optimum number of latent variables (after removing some y-outliers), with an optimum RMSECV of 0.9 °Bx. The resulting vector of PLS regression coefficients has the shape shown in Fig. 8.9. As expected, in the high-absorbance and saturated regions, the vector is highly uncertain. These regions should be removed from the data matrix before PLS model building.

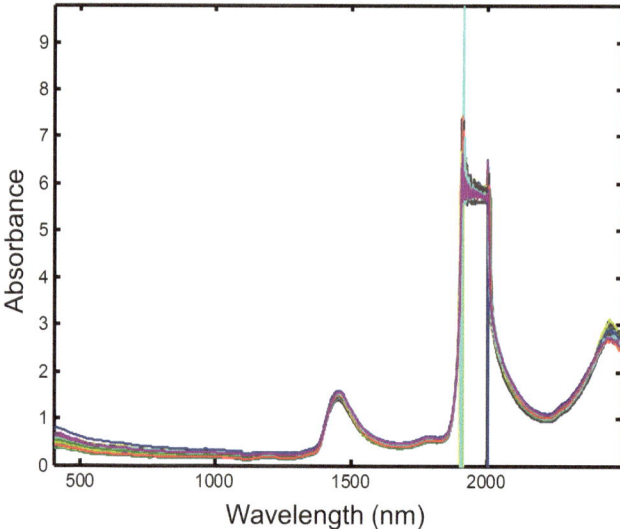

Fig. 8.8 Visible-NIR spectra for a set of sugarcane juice samples, employed to build a PLS model for the determination of the degrees Brix

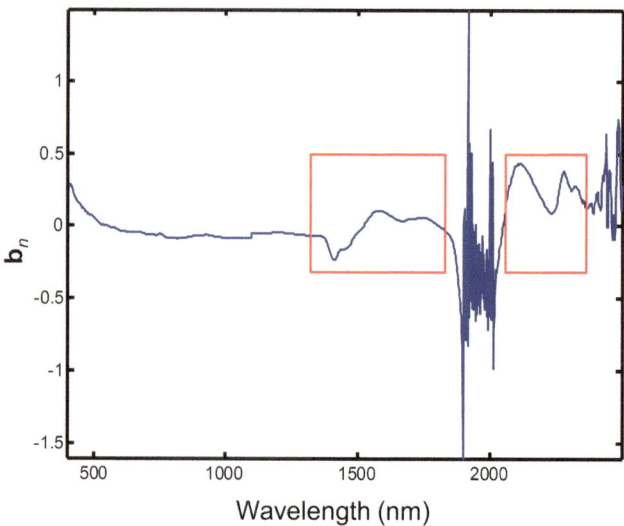

Fig. 8.9 Full-spectral PLS regression coefficients based on the spectra of Fig. 8.8 for the determination of degrees Brix of sugarcane juices. The red boxes indicate the sub-regions suggested for model building

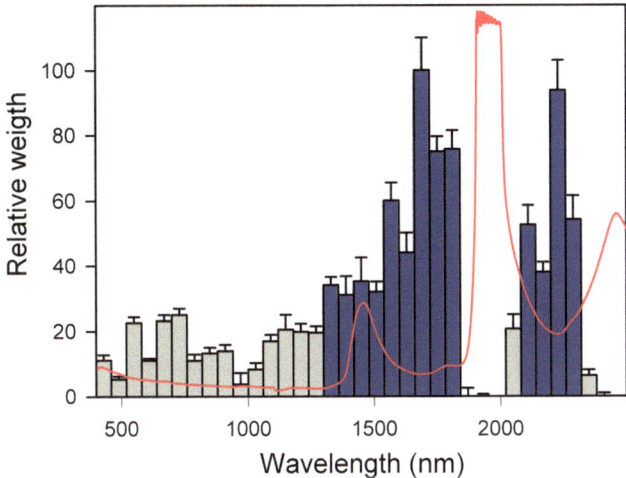

Fig. 8.10 Variable selection results applying a genetic algorithm to the data set of Fig. 8.8. The bars indicate the relative importance or weight of each sub-region, with standard deviations indicated on top of each bar, resulting from running the algorithm several times. The mean calibration spectrum is superimposed as a solid red line. Blue bars correspond to sub-regions included in the final model, while gray bars correspond to the excluded sub-regions. Adapted with permission from Sorol et al. (2010) (Elsevier)

   To improve the predictive power of the model, one would in principle select, by visual inspection, the regions marked with red boxes in Fig. 8.9. If this is done, only 5 latent variables are required by LOO cross validation (removing a single y-outlier), with RMSECV $= 0.26$ °Bx. Under these conditions, a set of 53 independent validation samples furnished an average prediction error (RMSEP) of 0.4 °Bx (2.4% with respect to the average value in the calibration samples of 17.9 °Bx). This is satisfactory for replacing the classical refractometric method by the visible-NIR/PLS model.

   In a published literature work on this same subject, complex algorithms for wavelength selection were applied (see next section), based on statistical/mathematical approaches which are beyond the scope of this book. The authors reached a final optimized model with RMSEP $= 0.3$ °Bx (1.6% of relative error) for the set of validation samples (Sorol et al. 2010). Not bad for our home-made method (RMSEP $= 0.4$ °Bx) based on the consideration of the vector of regression coefficients, right? In fact, the wavelengths selected using more sophisticated algorithms (Fig. 8.10) are close to those provided by our simple approach based on the properties of $\mathbf{b}_n$ (Fig. 8.9).

8.9   Other Sensor Selection Methods

Many additional procedures have been designed, which can be divided into:
(1) those employing the significance of the regression coefficients or loadings,
(2) those seeking the minimum prediction (or cross validation) error.

The first group includes, in addition to the visual inspection of the regression
coefficients, uninformative variable elimination (UVE) (Centner et al. 1996), which
focuses on the elements of the vector of regression coefficients and their significance, selecting those sensors for which the associated coefficients are larger than the
corresponding uncertainty.

The already discussed i-PCR/PLS method corresponds to the second group
(Nørgaard et al. 2000), as well as a number of probabilistic methodologies, which
combine individual sensors or groups of sensors, and estimate the prediction error
for cross validation or for a validation sample set. The combinations are
implemented following different mechanisms whose aim is to find a minimum of
the objective function (the prediction error), guided by models mimicking natural
processes. The latter may be, for example, the search of food by a colony of ants as in
ant colony optimization (Shamsipur et al. 2006), the motion of bird flocks following
a guide, as in particle swarm optimization (Xu et al. 2007), genetic algorithms
(Leardi and Lupiáñez González 1998), simulated annealing (Kalivas et al. 1989),
etc. Beyond the funny names of these algorithms, some of them are really effective
in the search for optimal spectral regions to be employed for a successful calibration.

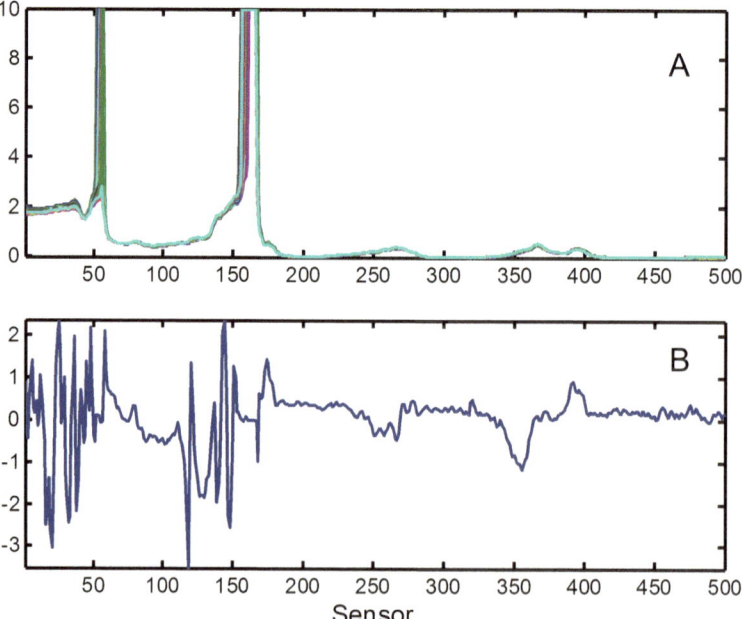

Fig. 8.11 (**a**) Spectra for a calibration set of gasoline samples, used to build a PLS model for the
determination of octane number. (**b**) Full-spectral PLS vector of regression coefficients

In the previously mentioned determination of degrees Brix in sugarcane juices, the authors applied various methods for wavelength selection; the most efficient ones were those inspired in natural computation (Sorol et al. 2010). Our simple method of inspecting the vector of regression coefficients, and the interval PLS procedure may yield excellent results for sensor selection, without resorting to seemingly complex methodologies.

8.10 Exercises

1. Figure 8.11a shows the NIR spectra of gasoline samples employed to build a PLS model for the determination of octane number, and Fig. 8.11b the corresponding full-spectral vector of regression coefficients.
   (a) Which sub-regions will you select for building a PLS model?
   (b) Explain the following results: for the full-spectral model, $A = 18$, RMSEP $= 0.56$, whereas after a suitable sensor selection, $A = 4$, RMSEP $= 0.27$.

References

Brown, C.D., Green, R.L.: Critical factors limiting the interpretation of regression vectors in multivariate calibration. Trends Anal. Chem. **28**, 506–514 (2009)

Centner, V., Massart, D.L., de Noord, O.E., de Jong, S., Vandeginste, B.M., Sterna, C.: Elimination of uninformative variables for multivariate calibration. Anal. Chem. **68**, 3851–3858 (1996)

Galvao, R.K.H., Araujo, M.C.U., Jose, G.E., Pontes, M.J.C., Silva, E.C., Saldanha, T.C.B.: A method for calibration and validation subset partitioning. Talanta. **67**, 736–740 (2005)

Kalivas, J.H., Roberts, N., Sutter, J.M.: Global optimization by simulated annealing with wavelength selection for ultraviolet–visible spectrophotometry. Anal. Chem. **61**, 2024–2030 (1989)

Kennard, W., Stone, L.A.: Computer aided design of experiments. Technometrics. **11**, 137–148 (1969)

Leardi, R., Lupiáñez González, A.: Genetic algorithms applied to feature selection in PLS regression: how and when to use them. Chemom. Intell. Lab. Syst. **41**, 195–207 (1998)

Mehmood, T., Liland, K.H., Snipen, L., Sæbø, S.: A review of variable selection methods in partial least squares regression. Chemom. Intell. Lab. Syst. **118**, 62–69 (2012)

Nørgaard, L., Saudland, A., Wagner, J., Nielsen, J.P., Munck, L., Engelsen, S.B.: Interval partial least-squares regression (iPLS): a comparative chemometric study with an example from near-infrared spectroscopy. Appl. Spectrosc. **54**, 413–419 (2000)

Shamsipur, M., Zare-Shahabadi, V., Hemmateenejad, B., Akhond, M.: Ant colony optimisation: a powerful tool for wavelength selection. J. Chemom. **20**, 146–157 (2006)

Sorol, N., Arancibia, E., Bortolato, S.A., Olivieri, A.C.: Visible/near infrared-partial least-squares analysis of Brix in sugar cane juice. A test field for variable selection methods. Chemom. Intell. Lab. Syst. **102**, 100–109 (2010)

Thomas, E.V., Haaland, D.M.: Partial least-squares methods for spectral analyses. 1. Relation to other quantitative calibration methods and the extraction of qualitative information. Anal. Chem. **60**, 1193–1202 (1988)

Xu, L., Jiang, J.H., Wu, H.L., Shen, G.L., Yu, R.Q.: Variable-weighted PLS. Chemom. Intell. Lab. Syst. **85**, 140–143 (2007)

Mathematical Pre-processing

9

Abstract

Multivariate calibration models sometimes require one to pre-process the instrumental data with mathematical techniques. Criteria are discussed for performing this relevant activity. The objective is to reduce the impact of physical phenomena or changes in the instrumental response over time.

9.1   Why Mathematical Pre-processing

NIR spectra, and other multivariate signals collected by means of reflectance measurements, i.e., after reflection on a solid or semi-solid sample, show background signals which vary from sample to sample, overlapped with the spectra of the chemical constituents. A typical example is observed in Fig. 9.1, where a high degree of dispersion is apparent, generating a baseline which significantly changes across samples.

These variable signals are produced by changes in the intensity dispersed by the sample, or by the presence of a variable optical path in the solid state or in powdered materials, and bear no relationship with the analyte concentration or chemical properties of a sample. They may be related to physical properties, as we shall see, but not to the chemical composition. A multivariate model built with spectra of the type of Fig. 9.1 will demand additional latent variables to those required by the chemical constituents, to be able to consider the dispersive effects. Is there any problem with this? In principle, the increment in latent variables conspires against the parsimony principle, and more parsimonious models are preferable, as previously discussed.

A variety of procedures has been developed, collectively called mathematical pre-processing, to reduce and, if possible, remove the effects of dispersion, and in general of any variable background signal. These procedures attempt to filter the

© Springer Nature Switzerland AG 2018   139
A. C. Olivieri, *Introduction to Multivariate Calibration*,
https://doi.org/10.1007/978-3-319-97097-4_9

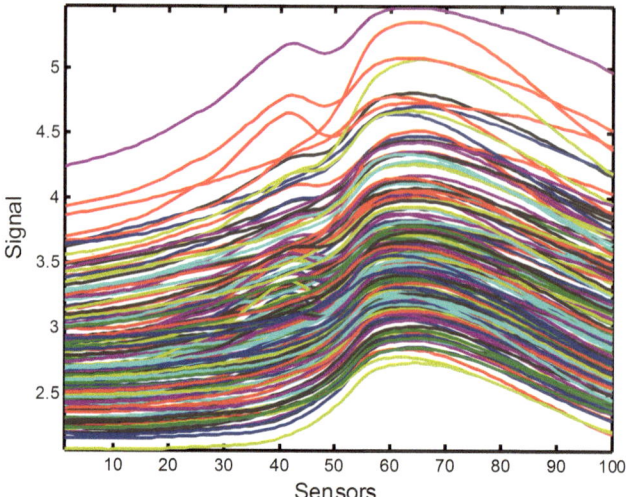

Fig. 9.1 NIR spectra of 170 meat samples, employed to build multivariate models to determine fat, moisture, and protein in a non-invasive manner. The data were recorded on a Tecator Infratec Food and Feed Analyzer working in the wavelength range 850–1050 nm, and are available at http://lib. stat.cmu.edu/datasets/tecator

signals, in such a way that the model is mainly concerned with the chemical composition of samples. Fewer latent variables will be expected in the latter case, leading to more parsimonious models.

The idea seems highly reasonable and has become very popular in multivariate calibration, particularly in the NIR spectroscopy field. However, these activities face at least two problems: (1) no rational recipe appears to exist for selecting the pre-processing method for a given system, and the issue becomes a question of trial and error, which may be time consuming and sub-optimal, and (2) some procedures, in particular those based on the concept of *moving window*, carry the risk of modifying the properties of the instrumental noise, introducing previously non-existing noise correlations, and making the models to be sub-optimal for analyte prediction.

A related activity is required when the measured signals vary over time or between different instruments: calibration maintenance and transfer, respectively. A suitable mathematical pre-processing method may restore the original predictive power of the calibration model or may allow one to transfer a calibration model from one instrument to another.

9.2   Mean Centering

An almost universally applied data pre-processing is mean centering. It consists in calculating the average calibration spectrum, and subtracting the latter from all spectra (calibration, validation, and unknowns). At the same time, the average

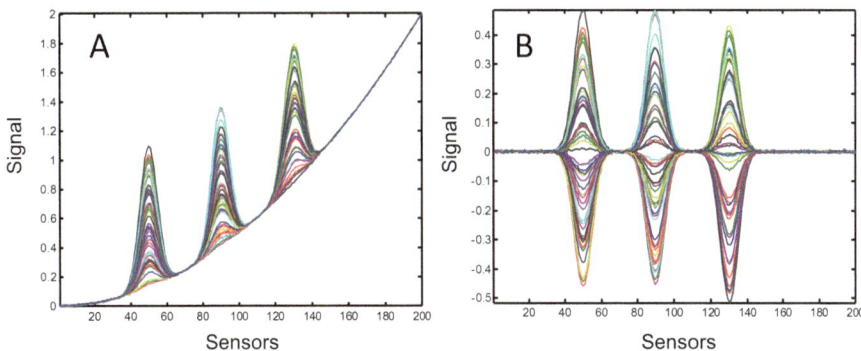

Fig. 9.2 (**a**) Calibration spectra of 50 samples containing three analytes and a constant background signal. (**b**) Mean-centered spectra

value of the calibration concentrations or properties is computed, and subtracted from all concentrations (calibration and validation). The model is then built with the centered spectra and concentrations, and the analyte concentration is predicted. This latter value will also be centered: it will be positive if larger than the calibration average and negative otherwise. After prediction, the analyte concentration is de-centered: the average calibration concentration is summed to the predicted value.

What do we get from mean centering? First, an almost formal achievement: to remove from all model expressions the intercept. Second, if a constant background signal occurs (equal for all samples), independently on the shape of this signal, mean centering will remove it. In this case, a centered model will require one less latent variable for calibration: the one that would be employed to model the background. Figure 9.2 shows the effect of mean centering on a data set with a constant background signal overlapped with the signals from three analytes: the background removal is apparent.

Unfortunately, dispersive background signals are seldom constant, and their general characteristic is that they vary from sample to sample (Fig. 9.1). Figure 9.3 shows the extreme case of a background signal strongly varying across samples (overlapped with the signals of three analytes), and the result after mean centering the spectra. As can be seen, the background persists, with positive and negative values. It is thus necessary to develop mathematical pre-processing procedures allowing one to correct for these effects. Some of them will be explained in the next sections.

9.3 Smoothing and Derivatives: Benefits and Hazards

Spectral smoothing and derivatives belong to the category of mathematical pre-processing employing the moving window strategy. They involve the selection of a certain window width (a small spectral region at the beginning of the spectrum),

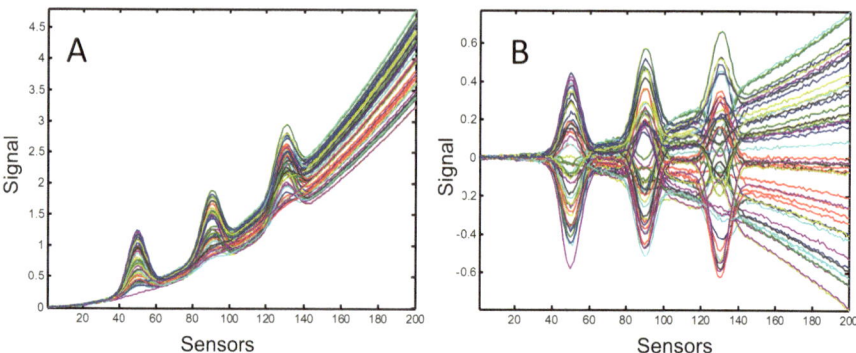

Fig. 9.3 (**a**) Calibration spectra of 50 samples containing three analytes and a variable background signal. (**b**) Mean-centered spectra

pre-process the data within the window, and then digitally move the window across the whole spectral range, sensor by sensor, applying the same mathematical pre-processing in each window.

The most employed tool for smoothing and derivatives is the Savitzky–Golay filter (Savitzky and Golay 1964). The report describing this technique represents a rare case of scientific success; in 2004 (40 years after its publication) it was one of the most cited papers in the prestigious *Analytical Chemistry* journal, published by the American Chemical Society. Some people consider the paper as starting the modern era of digital processing of signals. According to the database SCOPUS, in 2017 the paper had more than 8200 citations, a number that seems to have been exponentially growing in the last decade.

The basic Savitzky–Golay philosophy is simple: within a moving window of a certain width, the signal values are fitted to a polynomial function, estimating in each case the adjustable parameters (the coefficients of the polynomial terms). Subsequently, the signal corresponding to the center point of the window is estimated by the fitted polynomial (smoothing), or by the polynomial derivatives. The window is then moved across the spectral range and the activity is repeated. It is clear that the window width is conveniently set as an odd number, larger than the degree of the fitted polynomial, which should in turn be larger than the order of the derivative (for smoothing, the order is zero).

For example, suppose that the spectrum of Fig. 9.4, showing a considerable degree of random noise, requires: (1) smoothing and (2) estimating the first and second spectral derivatives. A third-degree polynomial is applied, with a five-sensor moving window. In the first window (Fig. 9.4), the signals are fitted to the following function:

$$y = ax^3 + bx^2 + cx + d \tag{9.1}$$

where a, b, c, and d are adjustable parameters, x represents each of the five sensors, and y the signal at a given sensor. With the fitted parameters, the value of y is

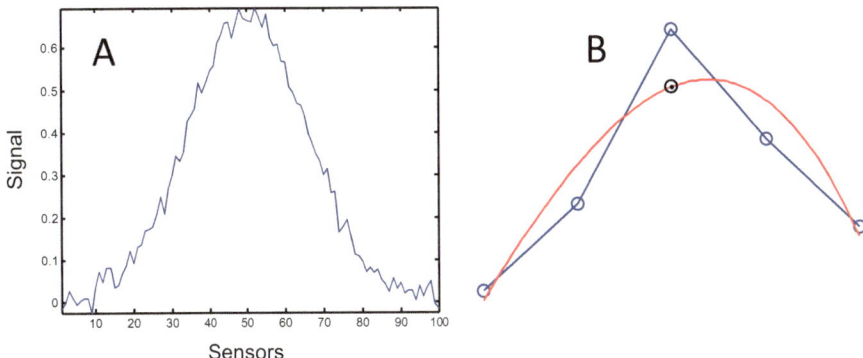

Fig. 9.4 (**a**) A spectrum with a high level of random noise. (**b**) The first 5-sensor window for the application of Savitzky–Golay smoothing (blue line and circles), a fitted third-degree polynomial (red line), and the value at the middle point estimated by the latter fit (black circle)

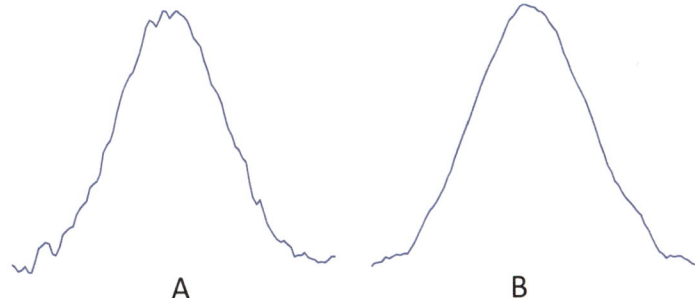

Fig. 9.5 Effect produced by smoothing as a function of the width of the moving window of the Savitzky–Golay filter on the spectrum of Fig. 9.4. (**a**) Using a 5-sensor window. (**b**) Using an 11-sensor window

estimated at the window center; this will be the first value of the smoothed spectrum. The process continues by moving the window, in such a way that the smoothed spectrum will have four less points than the original spectrum. In general, the smoothed spectra will end with a smaller number of points than the original; the difference is one less unit than the window width.

Having the fitted polynomial parameters, one can also estimate the derivatives. In each window, the following values for first and second derivative are estimated from Eq. (9.1):

$$dy/dx = 3ax^2 + 2bx + c \quad (\text{1st.derivative}) \tag{9.2}$$

$$d^2y/dx^2 = 6ax + 2b \quad (\text{2nd.derivative}) \tag{9.3}$$

The smoothing effect strongly depends on the size of the window. Figure 9.5 shows how the spectral smoothing progresses as a function of the window width. It is apparent that a high degree of smoothing can be achieved by applying the Savitzky–Golay filter, in this particular case, with an 11-sensor window. For illustrative purposes, or for the sake of preparing publication figures, the result could not be better. However, as previously anticipated, the smoothing should be considered as a spectral cosmetic, leading to a *seeming* improvement in the quality of the signal. Sadly, the random noise *is always there*. Is it beneficial to smooth spectra before building a multivariate calibration model? The apparent answer would be yes, but careful studies indicate otherwise: models based on digitally smoothed spectra by moving-window strategies may in fact be worse than those built with raw data (Brown and Wentzell 1999). In this context, notice the title of Brown and Wentzell (1999): *Hazards of digital smoothing filter as pre-processing in multivariate calibration*. The reason lies in the fact that these filters introduce correlations in the structure of the instrumental noise, with a range equal to the window size. This correlation, inexistent in the raw data, cannot be adequately modeled by PCR or PLS, which are based on the iid assumption for the instrumental noise (uncorrelated and with constant variance). The result is that PCR/PLS models based on smoothed data can be worse than those based on raw data (Brown & Wentzell, 1999).

Spectral derivatives estimated by the Savitzky–Golay method generate similar correlation effects in the noise, with an additional danger: the models may become sensitive to changes in the instrument reproducibility for wavelength registration. Small changes of a fraction of a wavelength may degrade the model performance.

The first derivative removes the effect of baselines varying across samples, something which cannot be achieved by mean centering (see previous section), provided the background signal is approximately linear (reader: can you explain why?). The second derivative is more general in this respect, and should be able to remove non-linear background signals varying from sample to sample, provided the non-linearity can be approximated by a quadratic function (reader: can you explain why?). We can appreciate the effect of the derivatives in Fig. 9.6, in comparison with Fig. 9.4, where mean centering was not effective in removing the background.

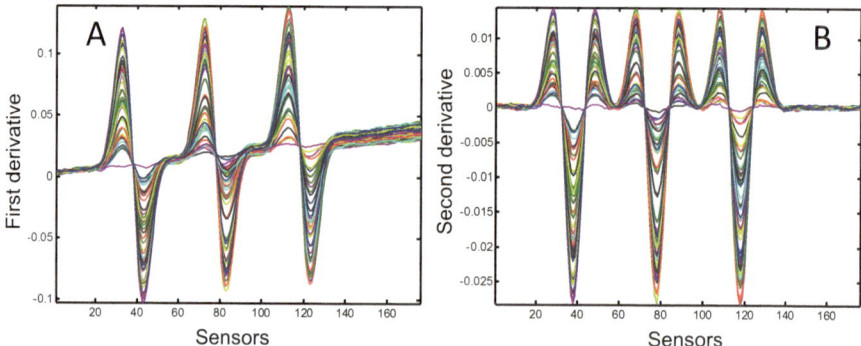

Fig. 9.6 (a) First derivative of the spectra in Fig. 9.3 estimated by the Savitzky–Golay filter. (b) Second derivative of the spectra in Fig. 9.3 estimated by the same filter. In both cases, a fifth-degree polynomial was used with a 25-sensor window

Figure 9.6 shows that the first derivative leaves some remaining baseline signal, but the second derivative has made a perfect job, because in this particular example, the baseline is approximately quadratic. Increasing the order of the derivative to cope with further non-linear backgrounds is not recommended. Spectral derivation increases the impact of the noise relative to the signal, and further smoothing the spectra may be even worse. Some researchers consider the second derivative as the universal method for removing undesirable dispersive effects.

9.4   An Algorithm for Smoothing and Derivatives

A simple MATLAB code to apply smoothing and derivatives is provided in Box 9.1. It estimates the coefficients for generalized values of window size and polynomial degree. Coefficient tables exist allowing one to directly estimate the smoothed spectrum and the derivatives from the signal values at each window point. Using these tables, the processed values are directly estimated as a linear combination of the signals at each window point. The tabulated coefficients are the weights of these linear combinations (Gorry, 1990).

Box 9.1

This algorithm implements smoothing and spectral derivatives. The variables present in the workspace should be "x" (the sample spectrum to be processed), "order" (the derivative order, with 0 implying smoothing, 1 first derivative and 2 second derivative), "degree" (the degree of the polynomial function, up to 5), and "window" (the width of the window, given by an odd number of sensors). The processed spectrum is stored in the variable "xnew":

```
J=size(x,1);
xnew=zeros(J-window+1);
v=1:window;
me=(window+1)/2;
mat=[v.^0;v.^1;v.^2;v.^3;v.^4;v.^5];
p0=[me.^0;me.^1;me.^2;me.^3;me.^4;me.^5];
p1=[0;1;2*me.^1;3*me.^2;4*me.^3;5*me.^4];
p2=[0;0;2;6*me.^1;12*me.^2;20*me.^3];
z=zeros(J,1);
for i=1:J-(window-1)
    data=x(i:i+window-1);
    coef=data'*pinv(mat(1:degree+1,:));
    eval(['z(i+me-1)=coef*p',int2str(order),'(1:',int2str(degree+1),');'])
end
xnew=z(me:J-me+1);
```

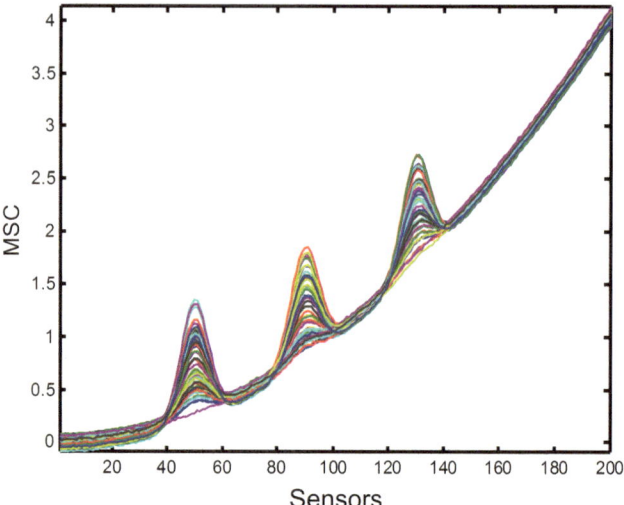

Fig. 9.7 MSC pre-processed spectra of Fig. 9.3

9.5 Multiplicative Scattering Correction

This mathematical pre-processing procedure, known by its acronym MSC, does not employ the moving window strategy, and does not present the noise correlation effect of the Savitzky–Golay smoothing or derivative filters. It is based on the idea that the background signals due to the light scattering, although they may change across samples, can be proportional or approximately proportional to each other, or even shifted by a constant value to lower or higher signal intensities. With this concept in mind, MSC first linearly fits by least-squares the spectra (calibration, validation, and unknown) to the mean calibration spectrum, and then subtract from all spectra the signal estimated by the linear fit.

The effect of MSC pre-processing over the spectra of Fig. 9.3 can be observed in Fig. 9.7. The procedure has definitely highlighted the concentration changes due to the analytes, because the bands for three constituents are clearly seen at the corresponding wavelengths for the spectral maxima. A remaining background signal is still visible, although the changes of this signal across samples are less variable than the raw data.

9.6 Additional Pre-processing Methods

There are other commonly applied tools for correcting spectra for dispersive effects: standard normal variate (SNV) and DETREND (implying removal of trends). In the former, the average signal and standard deviation are estimated for each spectrum.

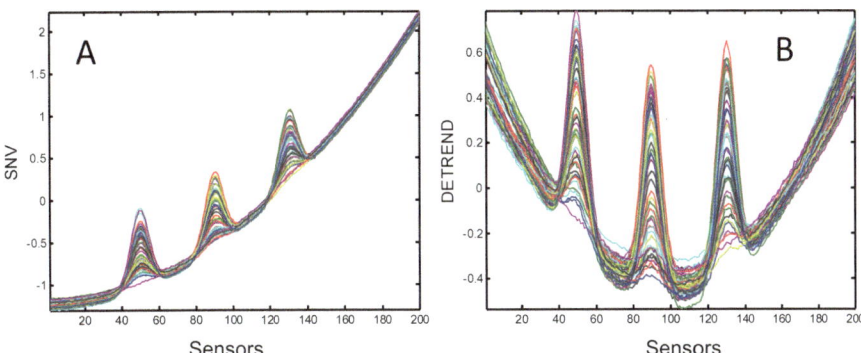

Fig. 9.8 (**a**) SNV pre-processed spectra of Fig. 9.3. (**b**) DETREND pre-processed spectra

The average signal is then subtracted from each spectrum, and the result is divided by the standard deviation. Figure 9.8a shows the effect of SNV on the spectra of Fig. 9.3. The result is similar to that achieved by MSC: there is a noticeable effect in highlighting the contribution of the three chemical constituents, particularly on the analyte centered at sensor 130, whose maximum signal is significantly overlapped with the background signal (Fig. 9.3b). However, some baseline signals still remain.

In DETREND, on the other hand, all the spectra are fitted by least-squares to a straight line, and the fitted values are then subtracted from each spectrum. This tool produces the effect shown in Fig. 9.8b, again highlighting the analyte effects, but leaving background signals.

A final detail concerns pre-processing methods which are not related to dispersion effects. In spectroscopic multivariate calibration, it is usually assumed that the spectra of a given constituent at unit concentration do not change from sample to sample. This is usually the case, so that the assumption is in general safe, but some multivariate signals do not follow this rule. Two examples are pertinent in this context: (1) chromatography and (2) electrochemical signals, e.g., voltammetry.

Liquid chromatographic experiments are well known for their intrinsic lack of reproducibility: if several aliquots of a given pure analyte solution are injected in a chromatograph, the measured traces at the detector will not overlap with each other. What can go wrong with a PLS model fed with chromatograms for a set of calibration mixtures? There are only a few studies in this respect, and they show that it is best to first synchronize the chromatograms (calibration, validation, and unknowns) with respect to a reference trace before model building and prediction.

In the determination of the enantiomers of ketoprofen by liquid chromatography using a chiral stationary phase, for example, a calibration set was designed with mixtures of the pure enantiomers (Padró et al. 2015). Figure 9.9a shows the raw calibration chromatograms, where the lack of synchronization is apparent. Digital displacement of the traces to match a reference chromatogram leads to Fig. 9.9b. The latter data were employed for successful PLS calibration and prediction.

In a different literature example, a PLS-1 model was developed based on differential pulse voltammetric traces measured on a glassy carbon electrode for the

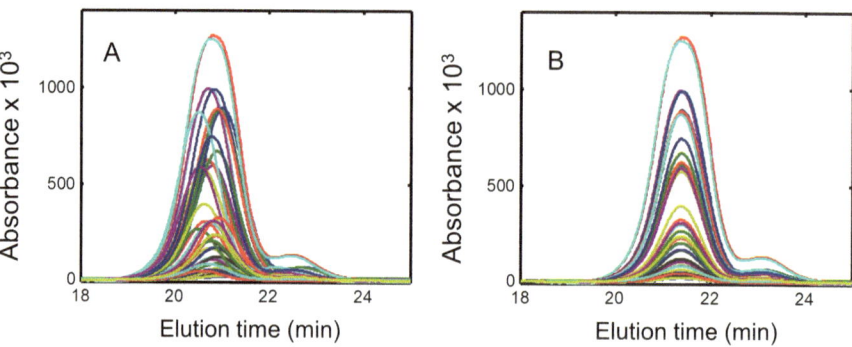

Fig. 9.9 (**a**) Raw liquid chromatograms of mixtures of (R) and (S)-ketoprofen. The signal is the absorbance at 220 nm. (**b**) The same chromatograms after temporal synchronization. Adapted with permission from Padró et al. (2015) (John Wiley and Sons)

simultaneous determination of the active principles levodopa, carbidopa, and benserazide in pharmaceutical formulations (Zapata-Urzúa et al. 2010). The authors showed that the peak potentials for the pure analytes varied with concentration, and thus first pre-processed the electrochemical profiles by synchronizing them using an efficient algorithm known as correlation optimized warping (COW) (Tomasi et al. 2004).

9.7   Algorithms for MSC, SNV, and DETREND

MATLAB codes for MSC, SNV, and DETREND can be found in Box 9.2.

Box 9.2
The following codes implement MSC. Here the variables are "Xcal" (the calibration data matrix) and "x" (the spectrum of a given sample, calibration, validation, or unknown). The code stores in variables "Xcalnew" and "xnew" the pre-processed calibration matrix and sample vector, respectively:

```
J=size(x,1);
espxcal=mean(Xcal,2)'-mean(Xcal(:));
espycal=Xcal-mean(Xcal(:));
beta=espycal'*pinv(espxcal);
alpha=(mean(Xcal)-beta'*sum(mean(Xcal,2)))/J;
espynew=x-mean(Xcal(:));
betanew=espynew'*pinv(espxcal);
alphanew=(mean(x)-betanew*sum(mean(Xcal,2)))/J;
Xcalnew=(Xcal-ones(J,1)*alpha)./(ones(J,1)*beta');
```

(continued)

Box 9.2 (continued)

```
xnew=(x-ones(J,1)*alphanew)./(ones(J,1)*betanew);
```

To apply SNV, the same variables are required as for MSC (Box 9.2), employing the following code:

```
[J,I]=size(Xcal);
for i=1:I
        sdelta=std(Xcal(:,i));
        Xcalnew(:,i)=(Xcal(:,i)-mean(Xcal(:,i)))/sdelta;
end
sdeltanew=std(x);
xnew=(x-mean(x))/sdeltanew;
```

DETREND uses the same variables as MSC (Box 9.2), and can be applied with:

```
J=size(Xcal,1);
mat=[ones(J,1),[1:J]'];
ab=Xcal'*pinv(mat');
Xcalnew=Xcal-mat*ab';
abnew=x'*pinv(mat');
xnew=x-mat*abnew';
```

9.8 How to Choose the Best Mathematical Pre-processing

A relevant question in pre-processing spectra concerns the existence of rules for the selection of procedures. Usually this important phase is conducted by a classical trial-and-error search. Attempts to find rational procedures for selecting the best mathematical pre-processing exist, including the careful analysis of the data structure (Brown 2000; Brown et al. 2000), or the use of experimental design and optimization (Gerretzen et al. 2015).

Most calibration developers still use the trial-and-error method. This consists in trying different procedures, separately or combined, until a good multivariate model is found, with acceptable average cross validation errors and number of latent variables.

An intermediate alternative is to select the pre-processing automatically, resorting to algorithms which are guided, as objective function, by the search of the minimum prediction or cross validation error (Devos and Duponchel 2011). Moreover, these algorithms can be combined with sample and wavelength selection, under the philosophy that pre-processing, samples, and sensors are an inseparable triad, and that they depend on each other, as depicted in Fig. 9.10. An algorithm of this type employs the following methods for optimizing a PLS model: (1) the Kennard–Stone method for selecting samples, (2) ant colony optimization for selecting wavelengths, and (3) a genetic algorithm for selecting math pre-processing (Allegrini and Olivieri 2013).

Fig. 9.10 Illustration of the
mutual relationship among
samples, sensors, and
mathematical pre-processing

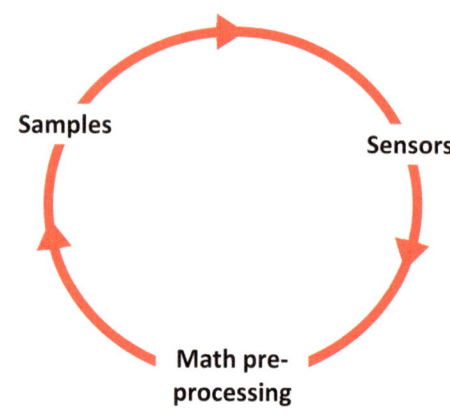

9.9   Is Pre-processing Always Useful?

As previously discussed, spectral pre-processing seeks to decrease or remove the
effect of physical phenomena varying from sample to sample (scattering, variable
optical path, etc.), in such a way that the model can focus with higher efficiency (less
latent variables, lower prediction errors) on the chemical composition of the studied
material. However, if the aim of the analysis is not chemical but physical, the
scattering may contain valuable information, and applying signal pre-processing
may seriously harm the model.

As an example, when trying to calibrate a multivariate model based on NIR
spectroscopy to measure the density and other physical properties of wood samples,
it was necessary to maintain the spectra in their raw form, not applying any
mathematical pre-processing as those presently discussed (Lestander et al. 2008).
Why? Because the NIR scattering is the phenomenon related to wood density!

On the other hand, liquid samples measured by transmission should not present
significant dispersion effects, so that spectral pre-processing would not be
recommended, at least in principle. It is usual to find publications where different
analytical results are compared for liquid samples, whose spectra were measured by
transmission, applying various pre-processing techniques. What we would expect in
these cases? Answer: small and marginal differences in the average validation errors,
most probably with no statistical significance. Pre-processing is not really needed!
An exception may be found in systems having constituents with highly overlapped
spectra: spectral derivatives are known to increase spectral resolution by reducing
the widths of constituent signals, which may lead to better prediction results in
comparison with raw data.

Table 9.1 Analytical results after different pre-processing methods are applied to the spectra of Fig. 9.3

Pre-processing	RMSECV	Latent variables
None	0.0054	4
Smoothing	0.0055	4
First derivative	0.0057	4
Second derivative	0.0056	3
MSC	0.071	2
DETREND	0.0055	3
SNV	0.072	2

9.10   A Simulated Example

What about the analytical predictions which can be expected after signal pre-processing? Table 9.1 shows the average cross validation errors (RMSECV) and optimum number of latent variables which are estimated for the set of spectra of Fig. 9.3. The analyte of interest is the one whose spectral maximum is located in the middle of the three constituent peaks. In all cases, the spectra were mean centered after each pre-processing was applied. For the Savitzky–Golay filters, a 15-sensor window and a fifth-degree polynomial were employed.

In this synthetic example, no significant improvement is achieved in the RMSECV in comparison with the use of raw data, but the optimum number of latent variables differs, implying that some PLS models are more parsimonious. The need of four variables when no pre-processing is applied is explained by the presence of the three constituents and the baseline signal (which leads PLS to demand an extra latent variable). This number is also four for smoothing and first derivative.

The second derivate, on the other hand, requires one less latent variable, which can be interpreted as a success of this procedure, because the pre-processed data carry information on three chemical constituents, without a significant effect from the background. DETREND leads to a similar result, without the danger of increasing the degree of correlation in the spectral noise.

Finally, in this particular example, MSC and SNV lead to only two latent variables, with an increase in the average cross validation error. However, this does not mean that they will not be useful in other examples.

9.11   A Real Case

In this section, we comment on the determination of total oil content in corn seed samples. The NIR spectra for the calibration samples are collected in Fig. 9.11., where a variable background signal is clearly visible. Pre-processing the spectra with first and second derivatives leads to the results shown in Fig. 9.12a, b, while MSC furnishes the spectra in Fig. 9.13. The result is surprisingly good: the derivatives leave almost no background signal. On the other hand, although MSC leaves a remaining baseline, it is almost constant, and could be removed by mean centering.

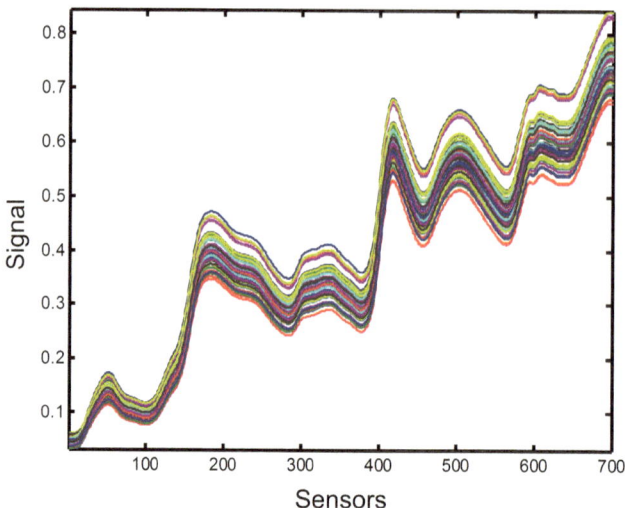

Fig. 9.11 NIR spectra of 50 corn seed samples, employed to build a model for the determination of quality parameters. They were measured in the wavelength range 1100–2498 nm at 2 nm intervals (700 channels), and are available at http://www.eigenvector.com/data/Corn

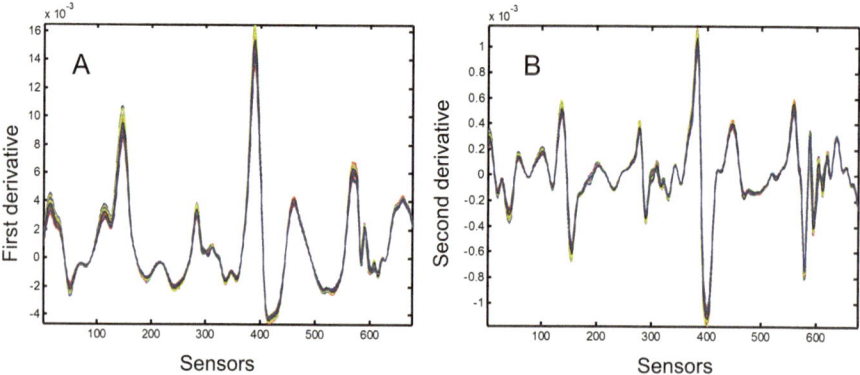

Fig. 9.12 (a) First derivative of the spectra of Fig. 9.11. (b) Second derivative. In both cases, a fifth-degree polynomial was used with a 25-sensor window

After pre-processing, the calibration spectra were submitted to PLS modeling, with the results summarized in Table 9.2., in terms of relative error of prediction (REP) on an independent validation set for various pre-processing methods and some combinations. The best models appear to be those highlighted in boldface in Table 9.2, i.e., first or second derivative and DETREND + derivatives. These models require a considerably smaller number of latent variables in comparison with the use of raw data, as expected from the removal of sample-to-sample variable background signals. The average validation errors are similar, but the more parsimonious models are to be preferred.

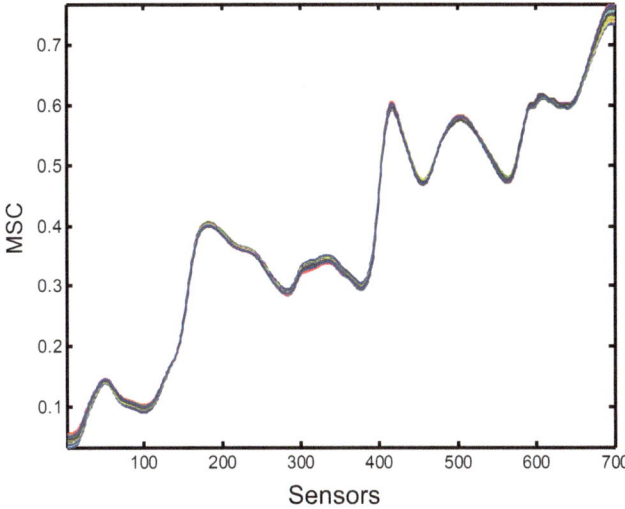

Fig. 9.13 MSC pre-processed spectra of Fig. 9.11

Table 9.2 Optimum number of latent variables and relative error of prediction (%) for different pre-processing methods applied to NIR spectra of corn seeds, employed to build PLS models for the measurement of total oil

Pre-processing[a]		Latent variables	REP (%)
None	No derivatives	21	1.1
	1st./2nd. derivative	**11**	**1.2**
MSC	No derivatives	19	2.0
	1st./2nd. derivative	10	2.0
SNV	No derivatives	19	1.9
	1st./2nd. derivative	11	2.1
DETREND	No derivatives	21	1.2
	1st./2nd. derivative	**11**	**1.1**
SNV	No derivatives	19	1.8
	1st./2nd. derivative	11	2.0

[a]In each case, the pre-processing in the first column is applied, followed by the one in the second column, followed by mean centering. The best models are highlighted in boldface

9.12 Calibration Update

We have already commented that multivariate calibration models may lose predictive power over time. Various examples were provided to illustrate a specific cause for this phenomenon: the appearance of constituents not present in the calibration set in new test samples. The universal solution to this problem was discussed as the need of expanding the calibration set to restore the predictive power, adding samples which are representative of the new constituents.

Model predictions can change for reasons beyond the appearance of new constituents in test samples. Another relevant cause is the variation in the instrument response over time, which may in turn be due to changes in the light source, the detector properties, or the accuracy of wavelength setting (Feudale et al. 2002). This needs to be monitored in time, for example, by checking whether the residuals of the modeling phase of the test sample signals maintain within a certain tolerated value. The control can be done with the F test discussed in Sect. 7.7 of Chap. 7. However, industrial analysts are somewhat suspicious of purely mathematical and statistical tests, relying on a more practical monitoring activity. They periodically measure the analyte concentration or target property in a set of validation samples using a reference technique, comparing the results with those predicted by the model. If acceptable values of the average prediction error (RMSEP) and relative error of prediction (REP) are produced, there is no need for updating.

How can models be updated? The simplest method is to correct the slope and intercept of a predicted vs. nominal plot for validation samples. This works properly under a limited set of circumstances, e.g., when a new constituent appears in test samples at a fixed concentration.

A better alternative is to mathematically transform the current instrument response to the one in a different condition (Feudale et al. 2002). Two different situations may occur: (1) the instrumental properties have changed so that the model is losing its original predicting ability or (2) we would like to transfer the available original calibration to a new instrument, whose response is not expected to be the same as the original one. In any case, transfer samples are required, measured on both conditions. These transfer samples may be some of the original calibration samples, if they are still available and are sufficiently stable. Otherwise, stable transfer samples, e.g., earth oxide glasses, can be used.

Piecewise direct standardization (PDS) is one of the oldest and faithful methods for calibration update, maintenance, and transfer (Wang et al. 1991). Suppose the original calibration of an instrument under conditions labeled as "1" was built with a calibration data matrix $\mathbf{X}_1$ of size $J \times I_1$ (J = number of wavelengths and I_1 = number of samples). In general, we do not want to re-measure all the I_1 calibration samples in the new condition "2," but only a smaller sub-set of I_2 samples ($I_2 \ll I_1$). The spectra for the I_2 samples under condition "2" produce a data matrix $\mathbf{X}_{s2}$ of size $J \times I_2$. It is obvious that the transfer sample sub-set should be sufficiently representative; the Kennard–Stone algorithm discussed in Sect. 8.2 of Chap. 8 is useful for the present purpose.

Once the data matrix $\mathbf{X}_{s2}$ has been measured, the corresponding columns of $\mathbf{X}_1$ for the same samples are selected to form matrix $\mathbf{X}_{s1}$. Now the problem is how to correlate $\mathbf{X}_{s1}$ with $\mathbf{X}_{s2}$, i.e., to find a transformation matrix $\mathbf{F}$ capable of converting spectra measured in condition "2" back to condition "1." A clever approach is to correlate each row of $\mathbf{X}_{s1}$ with nearby rows of $\mathbf{X}_{s2}$, or *pieces* of $\mathbf{X}_{s2}$, hence the name of the PDS procedure. Figure 9.14 (left) illustrates the concept: the pth row of $\mathbf{X}_{s1}$ ($\mathbf{x}_{1p}$) is correlated with the pth piece $\mathbf{X}_p$ of matrix $\mathbf{X}_{s2}$. How many pieces are

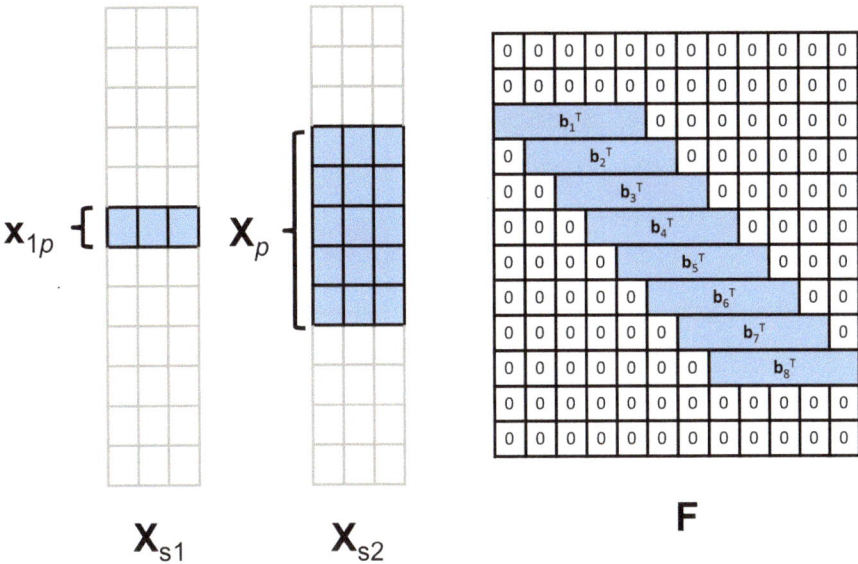

Fig. 9.14 Block illustration of the relevant PDS mathematical elements. Left: the *p*th row of the matrix $\mathbf{X}_{s1}$, designed as $\mathbf{x}_{1p}$, and a sub-matrix $\mathbf{X}_p$ of the matrix $\mathbf{X}_{s2}$. Right: the transformation $\mathbf{F}$ matrix. In both cases, 12 wavelengths were considered with a window with $K = 2$ sensors. In the transformation matrix, the $\mathbf{b}_p$ vectors are shown, for *p* running from 1 to 8

available? This depends on the size of the pieces: if the sensor windows before and after a given row $\mathbf{x}_{1p}$ are of size K, the piece size is $(2K + 1) \times I_2$, and there will be $P = J - 2K$ pieces. It is apparent that $2K$ sensors will be lost, K at the beginning and K at the end of the sensor range.

The expression needed to perform the required correlation (Fig. 9.14 left) is the following (transposition is made to bring it into a familiar form):

$$\mathbf{x}_{1p}^{\mathrm{T}} = \mathbf{X}_p^{\mathrm{T}} \mathbf{b}_p + \mathbf{e}^{\mathrm{T}} \tag{9.4}$$

where $\mathbf{b}_p$ is a vector of regression coefficients of size $(2K + 1) \times 1$. The reader will recall the ILS model equation (3.4) of Chap. 3, which is analogous to the present equation (9.4). Thus, each vector of regression coefficients $\mathbf{b}_p$ can be estimated by either PCR or PLS. Once all possible P rows of $\mathbf{X}_{s1}$ have been processed, a set of P vectors $\mathbf{b}_p$ are used to build the transformation matrix $\mathbf{F}$ of size $J \times J$. Figure 9.14 (right) illustrates the aspect of a typical $\mathbf{F}$ matrix: it has zeros outside a band of size $(2K + 1)$. Rows with zeros on the top and bottom account for the lost sensors.

Once $\mathbf{F}$ is available, it can be employed to transform any new vector measured in situation "2" ($\mathbf{x}_2$) back to situation "1" ($\mathbf{x}_1$):

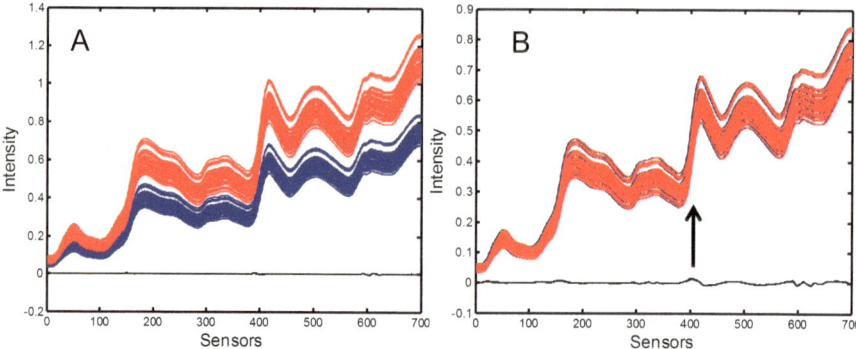

Fig. 9.15. (**a**) Blue, original NIR spectra for a set of 50 calibration samples, red, the same spectra multiplied by a constant factor of 1.5, black, residual spectra computed as the difference $(\mathbf{X}_1 - \mathbf{F}\mathbf{X}_2)$, with $\mathbf{F}$ estimated by PDS with a spectral window of six sensors and a sub-set of ten randomly selected samples. (**b**) Blue, original NIR spectra for a set of 50 calibration samples, red, the same spectra shifted two sensors to higher wavelengths, black, residual spectra computed as the difference $(\mathbf{X}_1 - \mathbf{F}\mathbf{X}_2)$, with $\mathbf{F}$ estimated as in (**a**). The black arrow in (**b**) indicates a spectral region where rather abrupt absorbance changes take place

$$\mathbf{x}_1 = \mathbf{F}\mathbf{x}_2 \qquad\qquad (9.5)$$

which could then be combined with the original vector of regression coefficients for the calibration in situation "1."

Figure 9.15 shows a simulation of two possible spectral changes: an increase in signal intensity by a constant factor at all sensors (Fig. 9.15a) and a shift in two sensors to higher wavelengths (Fig. 9.15b). The correction after applying the PDS procedure with a window of six sensors ($K = 3$) and a small randomly selected set of samples appears to be satisfactory, especially in Fig. 9.15a. It is apparent that wavelengths drifts are more challenging for correction (Fig. 9.15b).

9.13   A PDS Algorithm

A MATLAB code for PDS is provided in Box 9.3. The output of this short program is the transformation matrix needed to process the spectra of test samples, to bring them to the original conditions where calibration was built. The code requires one to set the value of the spectral window for the piecewise regression steps. This is usually done by trial and error, selecting the window value leading to the lowest average error for a set of validation samples.

Box 9.3

A PDS algorithm. The variables that should be present in the workspace are "Xs1," the matrix of sub-set of samples in condition "1," "Xs2," the matrix of sub-set of samples in condition "2," and "window," the number of channels for applying PDS.

```
J=size(Xs1,1);
F=zeros(J,J);
for j=window+1:J-window
    Xj=Xs2(j-window:j+window,:);
        [U,T,L]=princomp(Xj','econ');
    A=min(find(cumsum(L)/sum(L)>.99));
    bj=U(:,1:A)*pinv(T(:,1:A))*(Xs1(j,:)'-mean(Xs1(j,:)));
    F(j-window:j+window,j)=bj';
end
```

The output is the transformation matrix "**F**," which can be applied to transform a new spectrum "x," by the simple multiplication command F*x.

9.14 Exercises

1. Table 9.3 shows the results for the measurement of octane number in gasolines from NIR spectra, after application of various pre-processing methods (the minimum relative error is highlighted in boldface). Can you explain these results?

Table 9.3 Optimum number of latent variables and relative error of prediction (%) for different pre-processing methods applied to NIR spectra of gasolines, employed to build PLS models for the measurement of octane number

Pre-processing		Latent variables	REP (%)
None	No derivatives	4	0.27
	1st./2nd. derivative	4/5	0.37/0.46
MSC	**No derivatives**	**3**	**0.24**
	1st./2nd. derivative	5/5	0.34/0.47
SNV	No derivatives	4	0.28
	1st./2nd. derivative	4/4	0.38/0.47
DETREND	No derivatives	3	0.27
	1st./2nd. derivative	4/5	0.37/0.45
SNV	No derivatives	4	0.29
	1st./2nd. derivative	4/5	0.38/0.45

References

Allegrini, F., Olivieri, A.C.: An integrated approach to the simultaneous selection of variables, mathematical pre-processing and calibration samples in partial least-squares multivariate calibration. Talanta. **115**, 755–760 (2013)

Brown, C.D.: Ph. D. Thesis, University of Halifax, Canada (2000). http://dalspace.library.dal.ca/bitstream/handle/10222/55730/NQ60665.PDF?sequence=1.

Brown, C.D., Wentzell, P.D.: Hazards of digital smoothing filters as a preprocessing tool in multivariate calibration. J. Chemomet. **13**, 133–152 (1999)

Brown, C.D., Vega-Montoto, L., Wentzell, P.D.: Derivative preprocessing and optimal corrections for baseline drift in multivariate calibration. Appl. Spectrosc. **54**, 1055–1068 (2000)

Devos, O., Duponchel, L.: Parallel genetic algorithm co-optimization of spectral pre-processing and wavelength selection for PLS regression. Chemom. Intel. Lab. Syst. **107**, 50–58 (2011)

Feudale, R.N., Woody, N.A., Tan, H., Myles, A.J., Brown, S.D., Ferré, J.: Transfer of multivariate calibration models: a review. Chemom. Intel. Lab. Syst. **64**, 181–192 (2002)

Gerretzen, J., Szymańska, E., Jansen, J.J., Bart, J., van Manen, H.J., van den Heuvel, E.R., Buydens, L.M.C.: Simple and effective way for data preprocessing selection based on design of experiments. Anal. Chem. **87**, 12096–12103 (2015)

Gorry, A.: General least-squares smoothing and differentiation by the convolution (Savitzky–Golay) method. Anal. Chem. **62**, 570–573 (1990).

Lestander, T.A., Lindeberg, J., Eriksson, D., Bergstend, U.: Prediction of Pinus sylvestris clearwood properties using NIR spectroscopy and biorthogonal partial least squares regression. Can. J. For. Res. **38**, 2052–2062 (2008)

Padró, J.M., Osorio-Grisales, J., Arancibia, J.A., Olivieri, A.C., Castells, C.B.: Scope of partial least-squares regression applied to the enantiomeric composition determination of ketoprofen from strongly overlapped chromatographic profiles. J. Sep. Sci. **38**, 2423–2430 (2015)

Savitzky, A., Golay, M.J.E.: Smoothing and differentiation of data by simplified least squares procedures. Anal. Chem. **36**, 1627–1639 (1964)

Tomasi, G., Van den Berg, F., Andersson, C.: Correlation optimized warping and dynamic time warping as preprocessing methods for chromatographic data. J. Chemometr. **18**, 231–241 (2004)

Wang, Y., Veltkamp, D.J., Kowalski, B.R.: Multivariate instrument standardization. Anal. Chem. **63**, 2750–2756 (1991)

Zapata-Urzúa, C., Pérez-Ortiz, M., Bravo, M., Olivieri, A.C., Álvarez-Lueje, A.: Simultaneous voltammetric determination of levodopa, carbidopa and benserazide in pharmaceuticals using multivariate calibration. Talanta. **82**, 962–968 (2010)

Abstract

Figures of merit are regularly used to compare the performance of different analytical methodologies. A modern view of their definitions and interpretations is provided in the framework of first-order multivariate calibration.

10.1 Figures of Merit: What for?

A positive meaning is usually ascribed to the term *merit*. However, etymology indicates that the Latin word *meritum* should be translated as *the quality* of *being worthy of reward or punishment*, as the Oxford Dictionary does. Likewise, analytical figures of merit are quality parameters, which may favor or disfavor a certain methodology based on its efficiency (or lack of it).

The main usefulness of figures of merit lies in the possibility of comparing different analytical methodologies by means of simple, reliable, and easily interpretable numerical indicators. The analyst will usually balance the figures of merit, and other factors such as cost, operating time, possibility of automation, etc. before selecting a given analytical method for a specific application.

Some figures have an intrinsic importance that goes beyond the comparative analysis. For example, the limit of detection and the limit of quantitation are employed to establish whether a given analytical method can be applied or not to detect very low analyte concentrations. If an official protocol or norm establishes that the analytical methodology to be applied for the determination of a given analyte should be able to detect concentrations on the order of parts per billion, it is necessary to know the limit of detection to determine if the methodology is applicable or not at such low concentrations.

© Springer Nature Switzerland AG 2018   159
A. C. Olivieri, *Introduction to Multivariate Calibration*,
https://doi.org/10.1007/978-3-319-97097-4_10

10.2 Sensitivity

In classical univariate calibration, the sensitivity is defined by the International Union of Pure and Applied Chemistry (IUPAC) as the slope of the calibration graph (Fig. 10.1) (Currie 1995; Danzer and Currie 1998). The slope measures the change in signal for a unit change in concentration, which makes sense if we accept the qualitative definition of sensitivity as the response provoked by a certain stimulus. In analytical chemistry, the response is the instrumental signal and the stimulus is the analyte concentration. The definition can be extended to real life: a person who is easily offended or upset is said to be *sensitive*, whereas someone not aware of or able to respond to something is said to be *insensitive*. In all cases, the key is the proportion between stimulus and response.

An almost complete theory accounting for the sensitivity has been developed in the framework of multivariate and multi-way calibration. This theory defines the parameter in a different manner than above, based on error propagation theory and not on signal changes as a function of concentration (Olivieri 2014). Resorting to error propagation, if the analytical response is measured with an uncertainty s_x (assumed to be constant for all samples), and this uncertainty is propagated through the calibration process to all validation and unknown samples, a corresponding uncertainty s_y will be generated in the prediction of the analyte concentration. The numerical sensitivity parameter (SEN) is defined as the ratio between s_x and s_y:

$$\text{SEN} = s_x/s_y \tag{10.1}$$

and its units are signal $\times$ concentration^{-1}.

This definition is illustrated in Fig. 10.2. It is easy to show that Eq. (10.1) agrees with the IUPAC definition. We start from the prediction expression for the analyte concentration by means of the classical univariate graph:

$$y = (x - b)/a \tag{10.2}$$

where y is the predicted concentration, x the test sample signal, b the intercept, and a the slope of the calibration graph.

Fig. 10.1 A typical linear calibration graph, showing the slope (a) and intercept (b). The sensitivity is the value of the slope

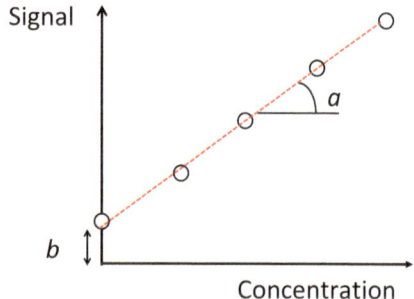

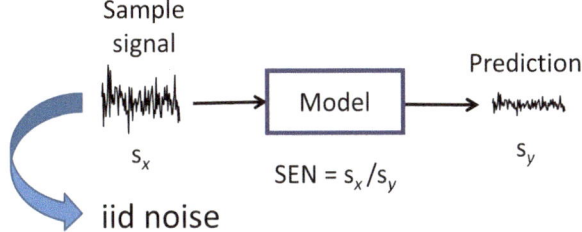

Fig. 10.2 Illustration of the estimation of the sensitivity, on the basis of error propagation of the signal for the test sample to the predicted concentration. The model is assumed to be precise, and the instrumental noise to be iid

We now assume that the calibration process is precise, meaning that the uncertainty only stems from the measurement of the test sample signal. If this is the case, the slope and intercept in Eq. (10.2) will carry a negligible uncertainty in comparison with the signal x, and a small error in signal (dx) will be transmitted to prediction as:

$$dy = dx/a \tag{10.3}$$

Multiplying each side of Eq. (10.3) by itself:

$$dy^2 = dx^2/a^2 \tag{10.4}$$

We should now consider that the square values dx^2 and dy^2 are averaged for a very large number of cases. The averages tend to the variances in x and y (s_x^2 and s_y^2), and their square roots to the standard errors s_x and s_y. Therefore:

$$s_y = s_x/a \tag{10.5}$$

The latter equation confirms that both sensitivity definitions agree, because from Eq. (10.5), SEN $= s_x/s_y = a$.

In first-order multivariate calibration, various sensitivity definitions have been proposed, but the accepted one today is based on error propagation (Fig. 10.2). In the inverse models PCR and PLS, we start from the prediction expression for the analyte concentration:

$$y = \mathbf{x}^T \mathbf{b}_n \tag{10.6}$$

It is then required to apply error propagation to Eq. (10.6), considering a small error in the signal at each sensor, and assuming that the elements of the vector of regression coefficients do not carry a significant uncertainty. The latter assumption is due to the fact that the calibration is assumed to precise. To simplify matters, let us consider an example where only two sensors exist; Equation (10.6) becomes:

$$y = x_1 b_{n1} + x_2 b_{n2} \tag{10.7}$$

Small changes in the signals at each sensor lead to the following change in concentration:

$$dy = dx_1 b_{n1} + dx_2 b_{n2} \tag{10.8}$$

As in the univariate definition, we multiply both sides of Eq. (10.8) by themselves:

$$dy^2 = dx_1^2 b_{n1}^2 + dx_2^2 b_{n2}^2 + 2 dx_1 dx_2 b_{n1} b_{n2} \tag{10.9}$$

What can be expected by averaging the values of dx_1^2, dx_2^2, and $(dx_1 dx_2)$ over a very large number of cases? The first two squares will become the variances at each sensor; if the instrumental noise is constant, the variances will both be equal to s_x^2. On the other hand, the cross product $(dx_1 dx_2)$ will tend to zero if there is no correlation between the noise at different sensors (sometimes the product $dx_1 dx_2$ is positive and sometimes it is negative). These two assumptions describe the characteristics of the so-called iid (identically and independently distributed) noise: constant variance and zero correlation.

Under this situation, Eq. (10.9) becomes:

$$s_y^2 = s_x^2 \left(b_{n1}^2 + b_{n2}^2 \right) \tag{10.10}$$

A convenient way of writing Eq. (10.10) is by realizing that the sum $\left(b_{n1}^2 + b_{n2}^2 \right)$ is the square of the length of the vector $\mathbf{b}_n$ (Pythagoras!). For a general case of J sensors, Eq. (10.10) can be written as:

$$s_y^2 = s_x^2 \sum_{j=1}^{J} b_{jn}^2 \tag{10.11}$$

From Eq. (10.11), it is easy to derive the multivariate sensitivity for PCR and PLS:

$$SEN = \frac{1}{\sqrt{\sum_{j=1}^{J} b_{jn}^2}} \tag{10.12}$$

Equation (10.12) shows that the sensitivity is the inverse of the length of the vector of regression coefficients. The units of SEN are (signal $\times$ concentration^{-1}), so that this parameter depends on the type of measured signal. As such, it is rather inefficient for comparing different analytical techniques, which may be based on wildly different signals. How to compare the sensitivity of a method given in spectral units of (absorbance M^{-1}) with an alternative one in electrical units of (mV M^{-1})?

To solve this problem, an additional figure of merit has been proposed: the analytical sensitivity γ, as the ratio between sensitivity and instrumental noise (Skogerboe and Grant 1970; Cuadros Rodríguez et al. 1993):

$$\gamma = \text{SEN}/s_x \tag{10.13}$$

This latter definition corresponds to univariate calibration, but can be extended to multivariate calibration in a simple way. The units of the parameter γ are inverse concentration, making it more useful for method comparison, because it does not depend on the specific type of signal.

Equations (10.12) and (10.13) may not be efficient for method comparison if the noise is not iid (Fragoso et al. 2016). Taking into account other noise structure is an advanced aspect of the present subject, beyond the scope of this book (Fragoso et al. 2016).

10.3   Selectivity

IUPAC mentions the selectivity in the framework of univariate calibration as *the extension that a method can be used to determine individual analytes in mixtures or matrices without interference from other components of similar behavior* (Vessman et al. 2001). In first-order multivariate calibration, as we have seen throughout this book, if the interferents are adequately represented in the calibration phase, they will not produce an interference in the classical sense. However, we may interpret the IUPAC definition in the multivariate world as indicating that the selectivity will be smaller than when the analyte is present in its pure form.

To represent the effect of the interferents on the selectivity, the latter has been defined as the ratio between the sensitivity parameter SEN of Eq. (10.12), and the sensitivity that the analyte would have in its pure form (SEN_0):

$$\text{SEL} = \text{SEN}/\text{SEN}_0 \tag{10.14}$$

With the latter definition, SEL is a dimensionless number varying between 0 and 1. Equation (10.14) can only be applied in the case the analyte is available in pure form, something which severely limits the usefulness of the expression, because in many applications the pure analyte is not available. For calibration of global sample properties (organoleptic, octane number, etc.), the selectivity parameter defined by Eq. (10.14) may not admit a viable chemical interpretation.

Instead of SEN_0 in Eq. (10.14), some researchers proposed the use of the total signal for a given sample at unit analyte concentration. With this definition, it will always be possible to estimate the selectivity, which will be a number smaller than 1. This definition leads to a selectivity parameter that varies from sample to sample, because the spectrum of the sample ($\mathbf{x}$) may display variable amounts of the interferents, in addition to the analyte. One expects that a figure of merit will qualify the calibration and will be independent on the test sample.

In sum, it is preferable not to report selectivity values for PCR or PLS calibrations.

10.4 Prediction Uncertainty

This is an important figure of merit, as revealed by the following statement of a well-known researcher: *a result without its estimated uncertainty cannot be taken seriously* (De Bièvre 1997). We have already commented on the uncertainty in the predicted analyte concentration in connection with the sensitivity, where the calibration parameters were considered to be infinitely precise, implying that the only source of uncertainty is the experimental collection of the spectrum for the unknown sample.

However, estimating the total prediction uncertainty requires one to consider the propagation of instrumental errors from the calibration phase, as well as the errors from the preparation of the calibration samples, or from the measurement of the reference nominal values for the calibration samples. The best literature source in this context is the classical work of Faber and Kowalski (of almost 60 pages long!), containing everything you wanted to know about uncertainty but were afraid to ask (Faber and Kowalski 1997). The main result is that the global prediction uncertainty can be explained by the propagation of three error sources: (1) the measurement of the multivariate signal for the unknown sample, whose variance is assumed to be constant and independent (the iid approximation) and equal to s_x^2, (2) the measurement of the signals for the calibration samples, having the same variance s_x^2, and (3) the analyte concentrations or properties for the calibration samples, which are affected by a constant and independent variance given by s_{ycal}^2, which is a function of the manner in which the calibration samples were prepared from the pure analyte and the potential interferents, or on how the reference calibration values were measured.

The prediction variance will then be equal to the sum of the three variances, one for each of the three error sources mentioned above, but the relative impact of these errors is not the same. The signal errors for the test sample are propagated to the estimated concentration as inversely proportional to the sensitivity, whereas the remaining two sources, arising from the calibration phase, are scaled by a dimensionless parameter known as the sample leverage, and symbolized as h (Faber and Kowalski 1997):

$$s_y^2 = s_x^2/\mathrm{SEN}^2 + hs_x^2/\mathrm{SEN}^2 + hs_{ycal}^2 \tag{10.15}$$

The first term in the right-hand side comes from the combination of Eqs. (10.12) and (10.13), and represents the contribution from the test sample. The second term is analogous to the first one, but is scaled by the sample leverage h, and represents the contribution of the calibration signal errors. For test samples close to the calibration center, h is smaller than 1, illustrating the averaging effect of the use of multiple calibration samples. The third term measures the effect of the calibration concentration errors, and is also scaled by the leverage.

Mathematically, the sample leverage is calculated from the truncated calibration score matrix $\mathbf{T}_A$ and the corresponding score vector for the unknown sample $\mathbf{t}_A$, as the squared length of the sample leverage vector $\mathbf{h}$:

$$\mathbf{h} = \mathbf{t}_A^T \mathbf{T}_A^+ \quad h = \sum_{a=1}^{A} h_a^2 \tag{10.16}$$

where $\mathbf{T}_A^+$ is the generalized inverse of $\mathbf{T}_A$ and h_a is a generic element of $\mathbf{h}$. The leverage in Eq. (10.16) can be qualitatively interpreted as *sample score divided by calibration scores*. The official interpretation is that the value of h locates the unknown sample in the calibration space, relative to the center of the space if the data were previously mean centered. Samples with a small leverage are close to the calibration center, and better represented by the calibration, leading to comparatively smaller prediction errors. Conversely, high-leverage samples, located far away from the center or close to the calibration borders, will lead to relatively larger prediction errors. This behavior is analogous to classical univariate calibration.

How to estimate the variance ingredients needed in Eq. (10.15)? The value of s_x can be experimentally estimated by studying replicate spectra for blank samples, if they are available, or for typical calibration samples if they are not. The standard deviation of replicate spectral measurements, averaged over the whole sensor range, will provide an idea of the level of uncertainty in the signals. As regards the concentration uncertainty s_{ycal}, two different situations exist: (1) the samples are prepared in the laboratory, and thus s_{ycal} is the uncertainty associated to sample preparation, or (2) the nominal concentration values or sample properties are measured by a reference technique, and thus s_{ycal} is the uncertainty of the latter technique.

Alternatively, one may allow the model itself to estimate the uncertainties in both signal and concentration. The former can be taken as the average standard deviation for the spectral residuals, whereas the latter may be considered as the standard error of prediction, both during the cross validation process. In any case, these model-estimated uncertainties should be judiciously checked by the analyst for consistency with experience.

10.5   The Effect of Mathematical Pre-processing

Mean centering, as previously indicated, is a pre-processing method which is applied by default in PCR or PLS calibration. The spectra are centered by subtracting the average calibration spectrum, and the concentrations are centered by subtracting the average calibration concentration for the analyte of interest. When data are mean centered, the expression for prediction is:

$$y_n = (\mathbf{x} - \langle \mathbf{x} \rangle)^{\mathrm{T}} \mathbf{b}_n + \langle y \rangle \tag{10.17}$$

where $\langle \mathbf{x} \rangle$ is the average calibration spectrum and $\langle y \rangle$ the average calibration concentration.

From Eq. (10.17), it is apparent that the prediction variance needs two additional terms, corresponding to the variance of the average spectrum and concentration. Statistics shows that the variance of the average is equal to the variance divided by the number of averaged objects, in the present case the I calibration samples. This means that under mean centering, Eq. (10.15) should be corrected in the following way:

$$\begin{aligned} s_y^2 &= s_x^2/\mathrm{SEN}^2 + h s_x^2/\mathrm{SEN}^2 + h s_{\mathrm{ycal}}^2 + (s_x^2/I)/\mathrm{SEN}^2 + (s_{\mathrm{ycal}}^2/I) \\ &= s_x^2/\mathrm{SEN}^2 + h_{\mathrm{eff}} s_x^2/\mathrm{SEN}^2 + h_{\mathrm{eff}} s_{\mathrm{ycal}}^2 \end{aligned} \tag{10.18}$$

where the effective leverage h_{eff} has been defined as $(h + 1/I)$. Notice that if the number of calibration samples is large $(I > 100)$, the practical effect of the term $(1/I)$ may not be significant.

For the remaining pre-processing procedures, it is necessary to distinguish between those based on the moving window concept, as smoothing and derivatives, from methods such as MSC, SNV, and DETREND (see Chap. 9). Why? Because moving windows modify the noise structure, introducing correlations among the noise at different sensors, even when the original noise is of the iid type. In this case Eq. (10.18) is no longer valid, because it is based on the iid assumption for the noise structure. A discussion on the specific effect of the moving window filters on the prediction uncertainty is beyond the scope of this book, and is a subject of current chemometric research (Olivieri and Allegrini 2017).

MSC, SNV, and DETREND do not show this problem. However, one should take into account that for these pre-processing methods, the value of s_x in Eq. (10.18) will no longer be the standard deviation for the raw signal, but the one for the pre-processed signal. If this latter uncertainty is experimentally estimated from replicates, the spectra should be pre-processed before calculating the value of s_x. If, on the other hand, the uncertainty in signal is estimated by the model itself, s_x will be directly given by the model as the pre-processed signal uncertainty.

10.6   Detection Limit

The limit of detection is the minimum detectable concentration with a specified degree of confidence (Currie 1999). It is important to remark the last portion of the above definition: *with a specified degree of confidence*. This is due to the fact that two parameters exist, in principle, to establish the detectability of an analyte. One of them is the critical limit of decision limit (LC), a concentration value from which the analyte is declared to be present, but without the sufficient level of confidence. The second is the true limit of detection (LOD), from which the analyte can not only be

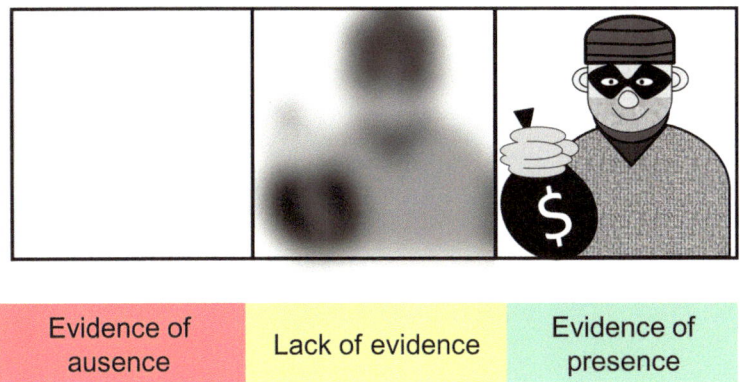

| Evidence of ausence | Lack of evidence | Evidence of presence |

Fig. 10.3 Analogy of decision and detection limits with real-life facts. Adapted with permission from (Olivieri and Escandar 2014) (Elsevier)

declared to be present, but its presence can be asserted *with a specified degree of confidence*. Notice the title of the publication in Faber (2008): *the limit of detection IS NOT the analyte level to decide between analyte detected and not detected*. This latter limit is, in fact, the critical limit LC.

The LOD is based on the existence of three concentration regions: (1) below the critical limit, (2) between the critical limit and the detection limit, and (3) above the detection limit. The definition is analogous to a well-known real-life situation (Fig. 10.3): (1) evidence of absence, (2) lack of evidence, and (3) evidence of presence (Olivieri and Escandar 2014).

To calculate the LOD value according to the above definition, it is required to estimate the standard error in the predicted analyte concentration for a blank sample (s_{y0}). This can be done by resorting to Equation (10.15), written in the following way (Currie 1999):

$$s_{y0}^2 = s_x^2/\text{SEN}^2 + h_0 s_x^2/\text{SEN}^2 + h_0 s_{ycal}^2 \tag{10.19}$$

where h_0 is the blank leverage, which is the leverage for a sample where the analyte is absent (recall that $1/I$ should be added if the data are mean centered).

As shown in Fig. 10.4, the LOD is estimated with a statistical hypothesis test. The first step is to set the concentration at the critical level (LC in Fig. 10.4), from which decisions are made with regard to analyte detection. For concentrations above the LC, there is a probability α of making a so-called Type I error, false positive or false detect. This latter error consists in erroneously accepting the alternative hypothesis, admitting that the analyte is present when it is in fact absent. As seen in Fig. 10.4, the relative probability of making Type I errors is given by the shadowed region at the right of LC (the area α). The distance from LC to zero concentration is approximately equal to the product of s_{y0} by the coefficient $t_{\alpha,\nu}$. If α is 0.05, a concentration higher than LC has a 5% probability to be a false positive. In the same vein, there is a probability β of making an error of Type II, false negative or false non-detect,

Fig. 10.4 Graphical representation of the significance test to estimate the limit of detection. The Gaussian curves indicate the concentration distributions at the blank level and at the limit of detection. The blue and red shadowed areas are the probabilities of making Types I and II errors, respectively

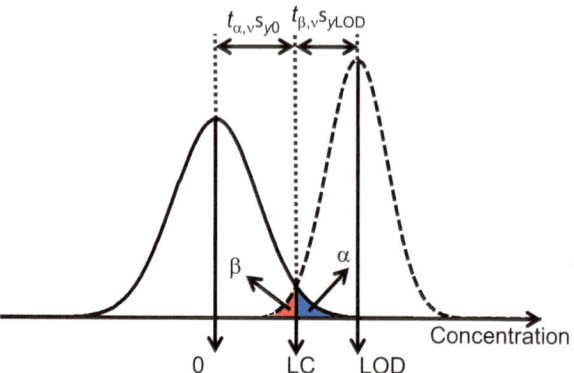

accepting the null hypothesis and admitting that the analyte is absent when it is in fact present (shadowed region at the left of LC in Fig. 10.4, with a probability β). If β is also taken as 0.05, the probability of obtaining a false negative result will be 5%. In this case the distance from LC to the concentration corresponding to this value of β is approximated by the product of the coefficient $t_{\beta,\nu}$ and s_{y0}. We are assuming that the standard error at the LOD level is very close to the standard error at the true blank level. The value of LOD thus depends on both α and β, and on the standard deviations of the two Gaussian curves in Fig. 10.4. If the standard deviations are assumed to be equal to s_{y0}, the LOD is given by:

$$LOD = 2t_{0.05,\nu}s_{y0} \tag{10.20}$$

In practice, since the number of degrees of freedom ν is large, the value of $(2t_{0.05,\nu})$ tends to 3.3, so that a good approximation to the limit of detection is:

$$LOD = 3.3s_{y0} \tag{10.21}$$

Notice that the LOD was formerly defined by only considering Type I errors, as the concentration corresponding to a signal/noise ratio equal to 3. This amounts to fixing the LOD as $LOD = 3s_{x0}/SEN$, where s_{x0} is the standard deviation at the blank level (in signal units). In this approximation, the probability of making Type I errors was of ca. 0.1%, corresponding to $t_{0.001,\nu} = 3$ (with large values of ν). This definition, today abandoned by IUPAC, does not consider Type II errors (Currie 1999).

The discussion on LOD would end at this point, with the complete expression for the limit of detection, as:

$$LOD = 3.3\left(s_x^2/SEN^2 + h_0 s_x^2/SEN^2 + h_0 s_{ycal}^2\right)^{1/2} \tag{10.22}$$

However, in Eq. (10.22) we need to estimate the blank leverage h_0, as discussed in the next section.

10.7   The Blank Leverage

In univariate calibration the leverage for a blank sample is estimated in a simple way, because in the absence of the analyte, the blank sample is unique. In multivariate calibration, on the other hand, there is no single blank. Because the analyte is determined in the presence of multiple interferents, well represented in the calibration samples, the blank sample may be a myriad of samples, all having variable amounts of the interferents and no analyte. Moreover, the blank might not exist at all: seeds without moisture, gasolines without octane number, etc. We refer, in general, to a virtual blank sample, not containing the analyte but containing the remaining constituents in variable concentrations.

We may conclude that the blanks are variable, and thus that there will be a range of blank leverages h_0. Can we estimate these values of h_0? Or put it in a different way, can we estimate the minimum and maximum blank leverages (h_{0min} and h_{0max})? If we could do it for a given PCR or PLS calibration, we could define a range of detection limits, from a minimum LOD_{min} to a maximum LOD_{max}, given by (Allegrini and Olivieri 2014):

$$LOD_{min} = 3.3 \left(s_x^2/SEN^2 + h_{0min}s_x^2/SEN^2 + h_{0min}s_{ycal}^2 \right)^{1/2} \tag{10.23}$$

$$LOD_{max} = 3.3 \left(s_x^2/SEN^2 + h_{0max}s_x^2/SEN^2 + h_{0max}s_{ycal}^2 \right)^{1/2} \tag{10.24}$$

The good news is that the extreme values of the blank leverage can be estimated in a relatively simple manner. Interestingly, the minimum value is identical to the one employed in univariate calibration, i.e., assuming that the only constituent in the samples is the analyte of interest (Allegrini and Olivieri 2014):

$$h_{0min} = \frac{\langle y \rangle^2}{\sum\limits_{i=1}^{I} (y_i - \langle y \rangle)^2} \tag{10.25}$$

where y_i is the analyte concentration in the ith calibration sample.

On the other hand, to estimate the maximum blank leverage it is first necessary to calculate the leverage shown by all calibration samples if the analyte could be virtually removed from them, projecting the actual leverages to zero analyte concentration:

$$h_{0i} = h_i + h_{0min} \left\{ 1 - [(y - \langle y \rangle)^2/\langle y \rangle^2] \right\} \tag{10.26}$$

where h_i is the leverage for each of the calibration samples and h_{0i} is the projected leverage mentioned above. The maximum of these projections is then selected (Allegrini and Olivieri 2014):

Zero analyte plane

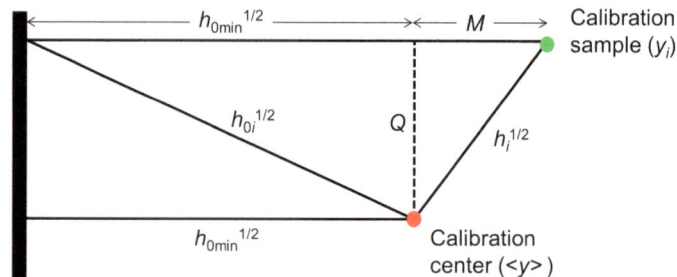

Fig. 10.5 Schematic representation of relevant sample leverages. The thick black line indicates the plane of zero analyte concentration, the red circle the location of the calibration center (analyte concentration $= \langle y \rangle$), and the green circle the location of the ith calibration sample (analyte concentration $= y_i$). Additional "distances" in score space (square roots of leverage values) are noted. Adapted from (Allegrini and Olivieri, 2014) (American Chemical Society)

$$h_{0max} = \max(h_{0i}) \qquad (10.27)$$

and can be used in Eq. (10.24) to estimate LOD_{max}.

Recall that for mean-centered data, $(1/I)$ should be added to h_{0min} and h_{0max}, to obtain effective leverages for insertion in Eqs. (10.23) and (10.24).

Interestingly, Eq. (10.26) can be derived from simple trigonometric arguments:

$$h_{0i} = h_{0min} + Q^2 = h_{0min} + (h_i - M^2) \qquad (10.28)$$

where the segments M and Q are defined in Fig. 10.5. From this figure, if the leverages are interpreted as squared distances proportional to concentration, then Eq. (10.26) immediately follows from Eq. (10.28).

The conclusion is that at zero analyte level, a range of sample leverages occur, which depend on the variability of the background composition, with two extreme values: the minimum (h_{0min}) given by Eq. (10.25), and the maximum of all h_{0i} values given by Eq. (10.26).

Figure 10.6 shows the leverages for a set of 20 calibration samples, all projected to zero analyte concentrations, for a typical multivariate system. The limits of detection corresponding to this system, presented in Fig. 10.7, do not show the same relative variability as the projected leverages of Fig. 10.6. This is due to the fact that the first term dominates the value of the LOD in Eqs. (10.23) and (10.24), whereas the ones scaled by the leverage show a comparatively smaller contribution. In this way, the LOD variability is attenuated with respect to the leverage variability.

As a final note, it is important to remark that the detection limit is useful in cases where future samples may have low or even null analyte concentrations, where it is really relevant to estimate low concentrations and quantitate the analyte detectability. Pertinent examples where it is important to detect the presence or absence of the analyte with a certain confidence are the determination of water in lyophilized

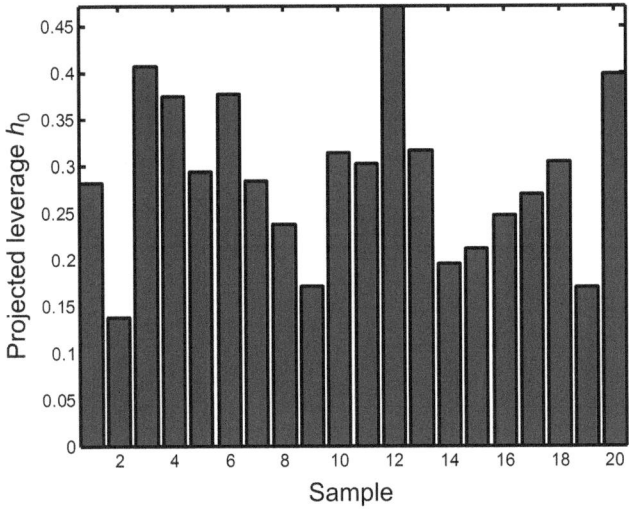

Fig. 10.6 Projections to zero analyte concentration of the leverages for 20 calibration samples of a typical multivariate system

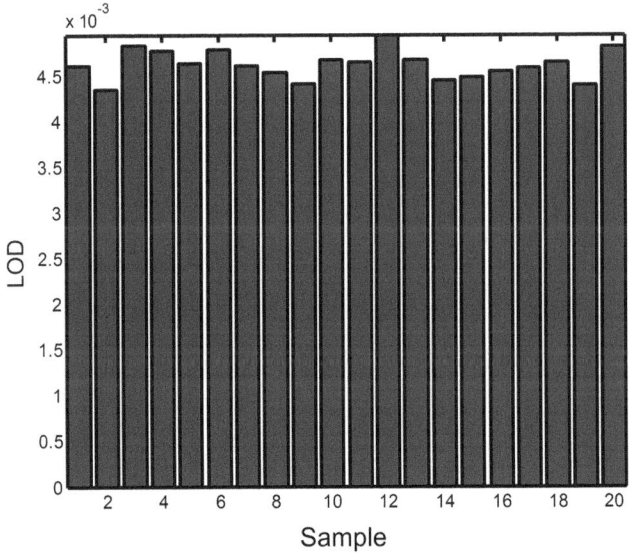

Fig. 10.7 Detection limits for the 20 calibration samples of Fig. 10.6

pharmaceuticals, adulterants in foodstuff, or therapeutic drugs in biological fluids. On the other hand, samples that will always contain high analyte concentrations may not require the LOD as a relevant figure of merit.

10.8   Quantitation Limit

The limit of quantitation (LOQ) is the analyte concentration for which the relative prediction error is at most 10%. The assumptions made in estimating LOQ are similar to those for the LOD, concerning the existence of a range of blank samples and blank leverages, and on the closeness of the LOQ value to zero analyte concentration. One could then estimate a minimum and a maximum LOQ value, analogously to the LOD discussion:

$$\mathrm{LOQ_{min}} = 10\left(s_x^2/\mathrm{SEN}^2 + h_{0\mathrm{min}}s_x^2/\mathrm{SEN}^2 + h_{0\mathrm{min}}s_{\mathrm{ycal}}^2\right)^{1/2} \qquad (10.29)$$

$$\mathrm{LOQ_{max}} = 10\left(s_x^2/\mathrm{SEN}^2 + h_{0\mathrm{max}}s_x^2/\mathrm{SEN}^2 + h_{0\mathrm{max}}s_{\mathrm{ycal}}^2\right)^{1/2} \qquad (10.30)$$

The factor 10 in Eqs. (10.29) and (10.30) is easy to explain: if the analyte concentration is ten times the prediction uncertainty, then the latter is 10% relative to the predicted concentration.

Figure 10.8 summarizes the concepts of limit of decision, limit of detection, and limit of quantitation.

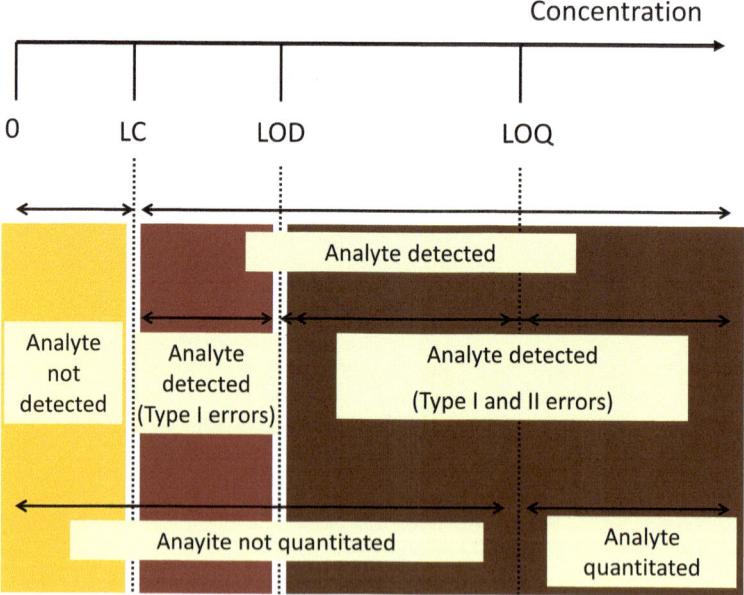

Fig. 10.8 Graphical illustration of the different concentration limits: decision (LC), detection (LOD), and quantitation (LOQ)

10.9   Other Figures of Merit

The figures of merit discussed above are those that may involve a certain conflict in their definition, and are not simply extrapolated from univariate to multivariate calibration. There are additional figures, such as accuracy, precision (including repeatability, intermediate precision, and reproducibility), robustness, etc. In the multivariate scenario, they deserve an analogous consideration as in univariate calibration.

10.10 Real Cases

In this section we first analyze an experimental case where the concept of limit of detection may be useful: the determination of the level of a fluorescent antibiotic in human plasma (Goicoechea and Olivieri 1999). The calibration phase involved a PLS model built from the synchronous fluorescence spectra of 50 human plasma samples, free from the presence of the antibiotic (tetracycline), spiked with the analyte in the concentration range from 0.0 to 4.0 ppm. In addition, 37 independent validation samples were prepared, as well as a final 20-sample set for the estimation of the limit of detection, containing low analyte concentrations (in the range 0.0–0.8 ppm). This latter experimental set was prepared because at the time the work was carried out, the LOD expressions (10.23) and (10.24) were not known, and the LOD was assessed experimentally. Today we can re-process these experimental data, and compare both LOD approximations.

Figure 10.9 shows the synchronous fluorescence spectra for the calibration samples, registered by scanning both excitation and emission monochromators of

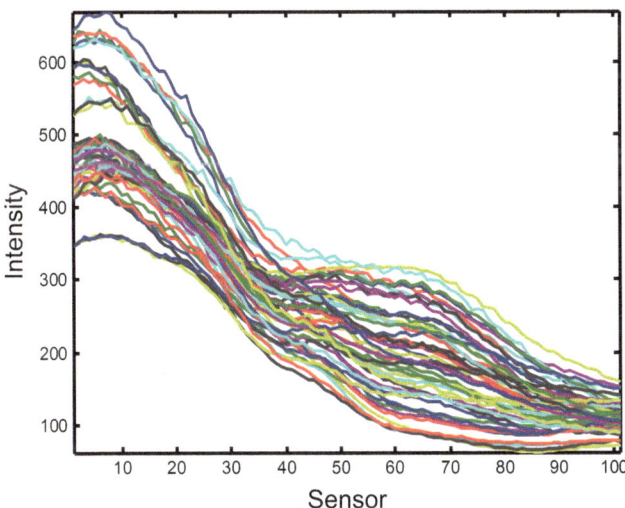

Fig. 10.9 Synchronous fluorescence spectra of 50 samples of human plasma with spiked tetracycline concentration in the range 0.0–4.0 ppm

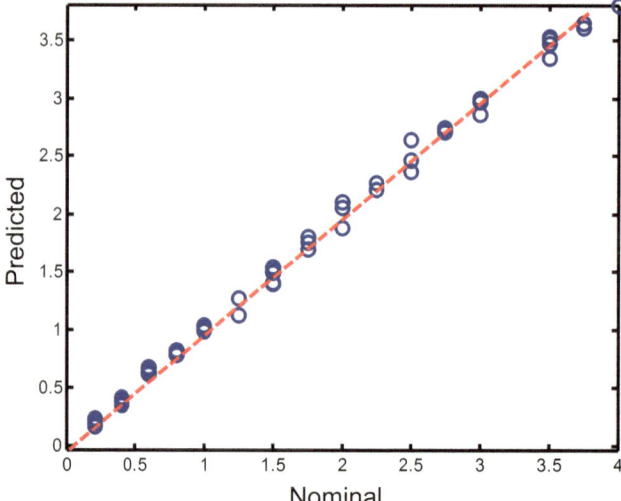

Fig. 10.10   Predicted vs. nominal tetracycline concentration for a set of validation plasma samples. The red dashed line has unit slope

the spectrofluorimeter, with a constant difference of 120 nm between them. This methodology allows one to obtain spectra with smaller bandwidths in comparison with classical spectrofluorimetry, increasing the spectral selectivity, and allowing better differentiation between the analyte and the interferents.

The PLS model was successful, leading to an RMSEP of 0.06 ppm (REP = 3.5%), after selection of the sensor range between 20 and 101 in Fig. 10.9 for calibration, and using three latent variables. The plot of predicted vs. nominal concentrations in the validation samples (Fig. 10.10) confirms the analytical indicators.

In the experimental approach to the LOD of Goicoechea and Olivieri (1999), the prediction standard errors of the analyte concentrations were computed in the low-concentration set of 20 samples, by means of the analysis of 4 replicates at each level. The LOD was established as the concentration whose average analyte concentration was larger than three times the standard error. The specific values at each level are reported in Table 10.1, with the result that LOD = 0.20 ppm (Goicoechea and Olivieri 1999).

In the theoretical approximation of Eqs. (10.23) and (10.24), on the other hand, the range of LOD values can now be estimated as 0.14–0.22 ppm, in excellent agreement with the experimental approach. The theoretical values do not require the preparation of the low-concentration 20-sample set, so today this activity would not be necessary for LOD estimation.

A complete report including analytical figures of merit for this system is presented in Table 10.2. It is important to notice that this report was prepared from the results offered by an appropriate computer program. In the next chapters, a MATLAB graphical user interface will be described which incorporates all the

Table 10.1 Nominal and predicted concentrations, and standard errors, for different levels of tetracycline in plasma at low concentration

Nominal (ppm)	Predicted (ppm)	Standard error (ppm)
0.0	0.00	0.02
0.2	0.20	0.03
0.4	0.38	0.03
0.6	0.65	0.03
0.8	0.81	0.02

Table 10.2 Complete report with figures of merit for the determination of tetracycline in human plasma

Analyte	Tetracycline
Samples	Spiked human plasma
Number of calibration samples	50
Number of validation samples	57
Analyte concentration range (ppm)	0–4
Optimum number of PLS latent variables	3
RMSECV (ppm)	0.13
RMSEP (ppm)	0.06
REP	3.5
SEN(fluorescence units $\times ppm^{-1}$)	103
γ (ppm^{-1})	32
LOD (ppm)	0.14–0.22
LOQ (ppm)	0.4–0.7

possibilities described previously for full-spectral inverse models such as PCR, PLS-1, and PLS-2, namely: (1) LOO and Monte Carlo cross validation, (2) outlier detection, (3) wavelength selection, (4) mathematical pre-processing methods, (5) analytical figures of merit, etc.

A second real example, worth mentioning here, is a recent work where the level of adulterants was assessed in saffron samples (Petrakis and Polissiou 2017). This is another situation where the limit of detection is relevant. Six typical plant-derived adulterants of saffron, namely *C. sativus* stamens, calendula, safflower, turmeric, buddleja, and gardenia were analyzed by PLS and diffuse reflectance infrared Fourier transform spectroscopy. For calibration and validation, 123 and 54 samples were employed, respectively. The limits of detection calculated with the approach described in Goicoechea and Olivieri (1999) were (in %): stamens, 2.2–3.1, calendula, 1.9–2.6, safflower, 2.1–2.8, turmeric, 1.0–1.6, buddleja, 1.1–1.6, and gardenia, 1.1–1.5 (Petrakis and Polissiou 2017). As can be seen, all LODs are provided as ranges, whose specific values allowed one to develop a rapid approach to saffron authentication.

10.11 Exercises

1. Indicate whether you agree or not with the report of Table 10.3.

Table 10.3 A report with figures of merit

Sample	Nominal	Prediction
1	0.10	0.0956(132)
2	0.20	0.2132(150)
3	0.30	0.2995(125)
4	0.40	0.4152(140)
5	0.50	0.5123(155)
RMSEP		0.0107
REP		35.8%
SEN		1.1589
LOD		0.01235
LOQ		0.03705

2. Which of the following statements is true? Explain.
   (a) A method with higher sensitivity should also have a lower limit of detection
   (b) A method with higher analytical sensitivity should also have a lower limit of detection
   (c) A method with a lower limit of detection should also have a lower limit of quantitation

References

Allegrini, F.A., Olivieri, A.C.: IUPAC-consistent approach to the limit of detection in partial least-squares calibration. Anal. Chem. **86**, 7858–7866 (2014)

Cuadros Rodríguez, L., García Campaña, A.M., Jiménez Linares, C., Román Ceba, M.: Estimation of performance characteristics of an analytical method using the data set of the calibration experiment. Anal. Lett. **26**, 1243–1258 (1993)

Currie, L.A.: Nomenclature in evaluation of analytical methods including detection and quantification capabilities. Pure Appl. Chem. **67**, 1699–1723 (1995)

Currie, L.A.: Detection and quantification limits: origins and historical overview. Anal. Chim. Acta. **391**, 127–134 (1999)

Danzer, K., Currie, L.A.: Guidelines for calibration in analytical chemistry. Part I. Fundamentals and single component calibration. Pure Appl. Chem. **70**, 993–1014 (1998)

De Bièvre, P.: Measurement results without statements of reliability should not be taken seriously. Accred. Qual. Assur. **2**, 269 (1997)

Faber, N.M.: The limit of detection is not the analyte level for deciding between "detected" and "not detected". Accred. Qual. Assur. **13**, 277–278 (2008)

Faber, K., Kowalski, B.R.: Propagation of measurement errors for the validation of predictions obtained by principal component regression and partial least squares. J. Chemometr. **11**, 181–238 (1997)

Fragoso, W., Allegrini, F., Olivieri, A.C.: A new and consistent parameter for measuring the quality of multivariate analytical methods: generalized analytical sensitivity. Anal. Chim. Acta. **933**, 43–49 (2016)

Goicoechea, H.C., Olivieri, A.C.: Enhanced synchronous spectrofluorometric determination of tetracycline in blood serum by chemometric analysis. Comparison of partial least-squares and hybrid linear analysis calibrations. Anal. Chem. **19**, 4361–4368 (1999)

Olivieri, A.C.: Analytical figures of merit: from univariate to multiway calibration. Chem. Rev. **114**, 5358–5378 (2014)

Olivieri, A.C., Allegrini, F.: Recent advances in analytical figures of merit: heteroscedasticity strikes back. Anal. Methods. **9**, 739–743 (2017)

Olivieri, A.C., Escandar, G.M.: Practical three-way calibration. Elsevier, Waltham (2014)

Petrakis, E.A., Polissiou, M.G.: Assessing saffron (Crocus sativus L.) adulteration with plant-derived adulterants by diffuse reflectance infrared Fourier transform spectroscopy coupled with chemometrics. Talanta. **162**, 558–566 (2017)

Skogerboe, R.K., Grant, C.L.: Comments on the definitions of the terms sensitivity and detection limits. Spectrosc. Lett. **3**, 215–220 (1970)

Vessman, J., Stefan, R.I., van Staden, J.F., Danzer, K., Lindner, W., Burns, D.T., Fajgelj, A., Müller, H.: Selectivity in analytical chemistry. Pure Appl. Chem. **73**, 1381–1386 (2001)

MVC1: Software for Multivariate Calibration **11**

Abstract

A simple, easy to use and intuitive software for first-order calibration is presented. It is freely available in the internet, and incorporates all the calibration models discussed in this book, including variable selection, pre-processing filters, and the latest advances in figures of merit.

11.1   Downloading and Installing the Software

The MVC1 software for first-order multivariate calibration was first described in the open literature in the form of a MATLAB graphical user interface (GUI) (Olivieri et al. 2004), as an improved version of an old Visual Basic program (Goicoechea and Olivieri 2000). The latest MATLAB version of MVC1 is freely available in the following site: www.iquir-conicet.gov.ar/descargas/mvc1.zip. On the other hand, a compiled, stand-alone version, which does not require to have MATLAB installed in the computer, is also available at: https://www.dropbox.com/sh/nruf3lpUge1gbww/AAAj6r97UBMIhgQmukRGYFPKa?dl=0.

Each version has pros and cons. The MATLAB codes can be implemented with the version R2012a, but may not be compatible with the latest MATLAB versions. They are open *.m files, which can be modified at will by skilled operators. The compiled version can be run without MATLAB, but operates as a black box, without access to the codes. To install the stand-alone version, a complementary file needs to be downloaded from the above site and installed. This file is known as MCR (MATLAB Common Runtime), and is freely distributed. Installation instructions, example data, and user manual are available with both MVC1 versions.

To run the compiled program, it is only necessary to execute it and select the data files with convenient browsers. In the case of the MATLAB version, it is first required to declare the program path, the folder containing the program codes,

© Springer Nature Switzerland AG 2018   179
A. C. Olivieri, *Introduction to Multivariate Calibration*,
https://doi.org/10.1007/978-3-319-97097-4_11

Table 11.1 Example MVC1 data sets

Data set	Analyte or property	Samples	Spectra
Bromhexine in syrups	Bromhexine	Cough syrups	UV-visible
I5 in reactor	A nitro-cresol	Industrial reaction mixtures	UV-visible
Tetracycline in serum	Tetracycline	Human plasma	Synchronous fluorescence
Octane in gasolines	Octane number	Gasolines	NIR
Parameters in corn	Oil, moisture, starch, and protein	Corn seeds	NIR
Parameters in meat	Fat, moisture, and protein	Meat	NIR

and the working folder, which is suggested to be different than the one containing the codes.

Independently on the selected version, it is recommended to periodically visit both internet sites, because MVC1 is being continually updated.

11.2   General Characteristics

Table 11.1 collects the example data accompanying MVC1, together with a brief description of each calibration system.[1] A good suggestion is to save the files for each data set in different folders.

There are some previous requirements to operate with the software, as described below.

1. A set of calibration spectra or multivariate signals is required to build the model. It is assumed that the signals were measured at J sensors for I samples, and can be organized in a data table or matrix, of size $J \times I$, and saved in a single flat text file. In MVC1, this data type is called "Matrix."

   The data could also be saved sample by sample, in separate text files, which may contain: (a) a vector of size $J \times 1$ of signal values, or (b) a two-column data table of size $J \times 2$, the first column being the wavelengths (or sensor values) and the second column the signal intensities. These data types are called, respectively, "X_vectors" and "X,Y_vectors."

   The analyst should know in advance the specific file type. Figure 11.1 shows a typical content of a "Matrix" file, and Fig. 11.2 those for "X_vectors" and "X,Y_vectors" files.

2. The calibration concentrations or nominal properties to be calibrated are also required. These values could be saved in: (a) a single one-column text file

---

[1]MVC1 includes an additional simulated data set with correlated noise, to be studied with the maximum likelihood MLPCR model. This subject is beyond the scope of this book.

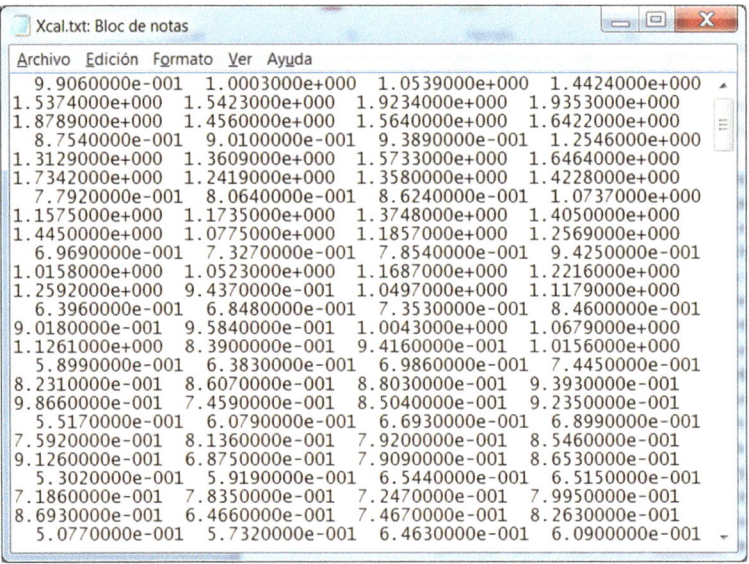

Fig. 11.1 Typical content of a text file of type "Matrix," showing the first four columns of the data table (each column is a spectrum and each row is a sample). Notice that the columns are separated by blank spaces

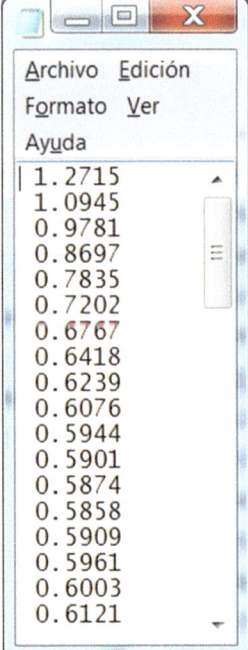 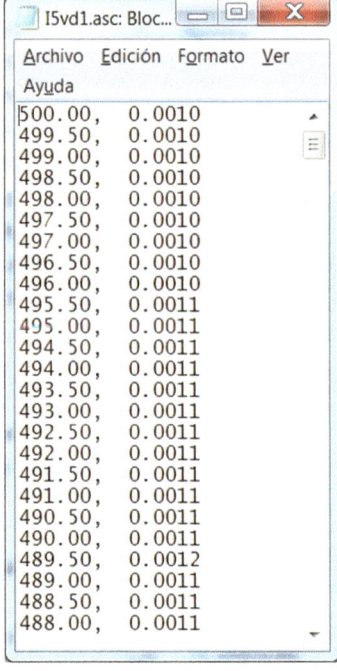

Fig. 11.2 Typical content of text files of types "X_vectors" and "X,Y_vectors" (left and right, respectively). In the former case, only the signal values were saved in the file; in the latter, the first column contains the wavelengths and the second one the signals. Notice that the columns in the left panel are separated by commas. Both blank spaces and commas are acceptable

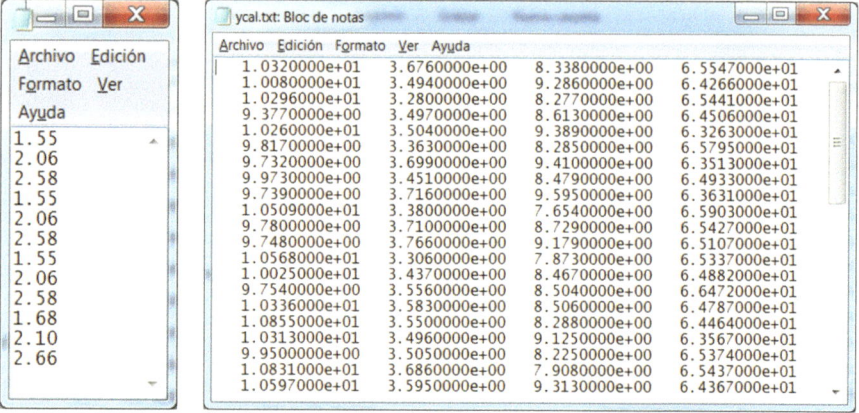

Fig. 11.3 Typical content of text files with calibration concentrations. Left: a single analyte. Right: multiple analytes. Each column corresponds to a given analyte and each row to a sample

(or several one-column files if there are more analytes or properties to calibrate) or (b) a single text file with as many columns as analytes or properties. Figure 11.3 shows typical concentration files for either one or several analytes.

3. If the intensity data are of "Matrix" type, then the first column of the data matrix will be associated to the first row of the concentration file, the second column of intensities with the second row of concentrations, and so on. Figure 11.4 illustrates the connection between signals and concentrations in this case.

4. If the data are of types "X_vectors" or "X,Y_vectors," an additional file will be required: the so-called *connecting* file between signals and concentrations. This connecting file must contain, one below each other, the names of the files with the signals of each calibration sample. It is clear that the order of the names in the connecting file must be the same as the order of the concentrations in the concentration file. Figure 11.5 shows a typical connecting file and its relationship with the concentration file.

5. Filenames should not start by a number, and should not contain blank spaces or mathematical symbols, such as "+," "*," "−," or "/". The underscore "_" character, however, is allowed.

6. For the prediction phase, it is usual to have some files available with similar characteristics to those for the calibration phase. For validation, the required files contain: (a) signals of type "Matrix," "X_vectors," or "X,Y_vectors," (b) validation concentrations (one or more columns, depending on the number of analytes), and (c) a connecting file relating validation signals and concentrations if the data are of types "X_vectors" or "X,Y_vectors."

   For truly unknown samples, only the signals will be available, with no information on nominal concentrations.

7. To run MVC1, execute the mvc1_gui.m routine in the MATLAB version and mvc1_32.exe in the compiled one.

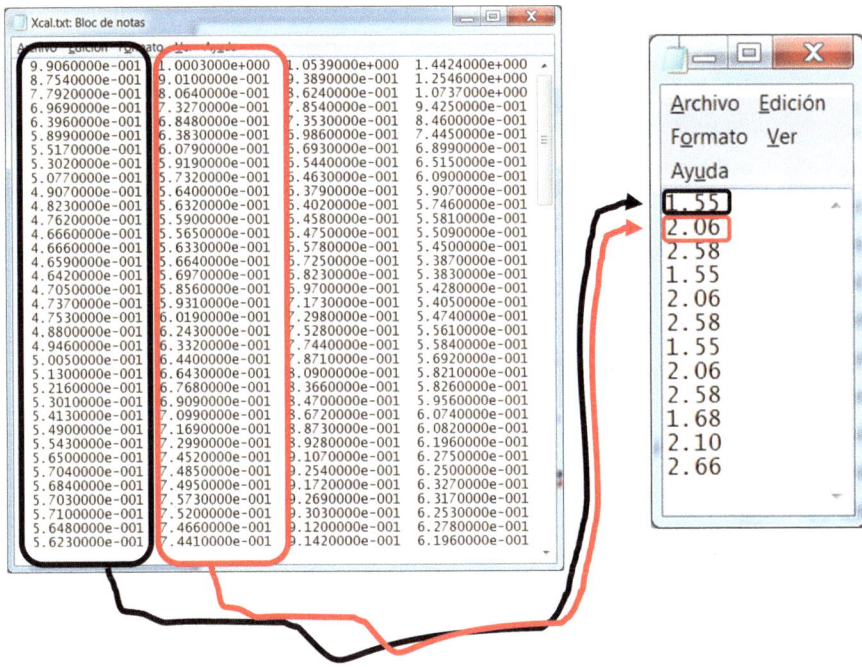

Fig. 11.4 Connection between columns of a file of type "Matrix," and rows of the calibration concentration file

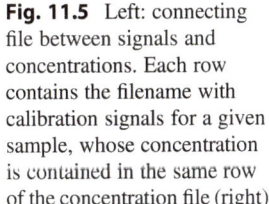

Fig. 11.5 Left: connecting file between signals and concentrations. Each row contains the filename with calibration signals for a given sample, whose concentration is contained in the same row of the concentration file (right)

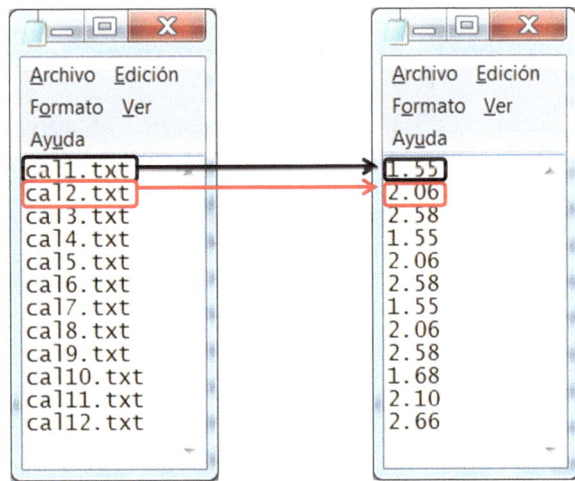

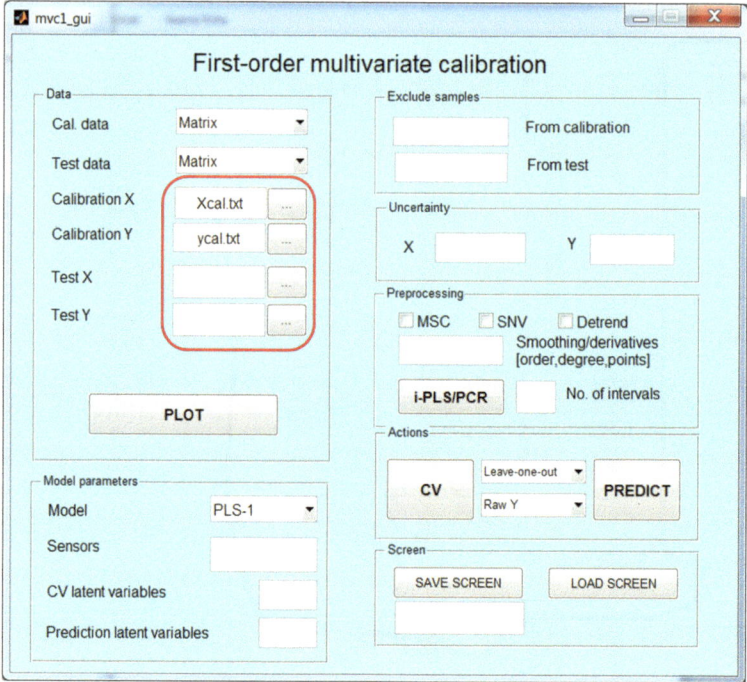

Fig. 11.6 Main MVC1 screen, showing in the red box the place reserved for loading the calibration files: signals ("Calibration X") and concentrations ("Calibration Y")

11.3 Real Cases Studied with MVC1

As detailed in Table 11.1, data for a series of analytical systems accompany the MVC1 software. All of them are experimental data sets, either measured in the author's laboratory (*Bromhexine in syrups, Tetracycline in serum*), in an industrial collaborating laboratory (*Octane in gasolines, I5 in reactor*) or available on the internet (*Parameters in corn, Parameters in meat*). Various spectroscopic techniques are involved in these determinations, namely UV-visible, synchronous fluorescence and NIR (both in solution and in solid samples).

Three of the above systems will be described in detail in the following sections: *Bromhexine in syrups, I5 in reactor*, and *Parameters in corn*. Two exercises in Sect. 11.10 ask the reader to process the data for *Tetracycline in serum* and *Octane number in gasolines*. The final system *Parameters in meat* is postponed for the next chapter, devoted to non-linear systems and artificial neural networks.

11.4 Bromhexine in Cough Syrups

This analysis was motivated by the interest of a pharmaceutical industry in replacing a classical liquid chromatographic method for the determination of the active principle of cough syrups by a more convenient spectroscopic method based on UV-visible measurements (Goicoechea and Olivieri 1999a). The active principle bromhexine absorbs in the UV-visible region, but its spectrum is overlapped with the remaining sample constituents, collectively known by the industry as the *syrup blank*. For calibration, 12 samples were prepared by mixing different levels of the syrup blank with different concentration of standard bromhexine.

The corresponding data folder which accompanies MVC1 shows that both "Matrix" and "X_vectors" types are available for this example. We choose to work with the former data type, meaning that after execution of the program, we should first provide the names of the files containing the calibration signals (Xcal.txt) and calibration concentrations (ycal.txt), using the MVC1 screen browsers (Fig. 11.6)

Some model names and other parameters appear in the MVC1 screen by default. If the analyst modifies the screen, and then writes a name in the blank space corresponding to "SAVE SCREEN" and presses the button, the new screen will be saved. In a future MVC1 session, it can be loaded using the "LOAD SCREEN" button (Fig. 11.6).

After pressing "PLOT" we can examine the calibration spectra. Figure 11.7 shows the raw data (left panel) and the pre-processed ones (right panel). In this case, no math filters have been selected, so that both panels look identical (recall that mean centering will be then applied by default in all models, but this is not shown in Fig. 11.7). The horizontal axis for the spectra of Fig. 11.7 does not provide information on wavelengths, but only on sensor indexes or data points.

In this particular example, validation data are available with spectra (Xtest.txt) and concentrations (ytest.txt), which can be loaded from the main screen, and plotted by means of the "PLOT" button (Fig. 11.8). Pressing "SAVE" allows one to save the pre-processed calibration and validation spectra in text files, in a folder called "temp" which is created in the working folder.

To build the calibration model, we need to first select the specific multivariate model to be applied, which can be done from the corresponding drop-down list in the main screen. In this case we select the option "PLS-1." We leave the space adjacent to "Sensors" empty, implying that the model will be built with all sensors or wavelengths. Otherwise, the space should be completed with the specific working sensors, as detailed below for other examples.

We then need to choose the maximum number of latent variables for performing cross validation; the default option is the LOO-type using the raw concentration values. Additional options will be explored below. This maximum number of latent variables should be roughly set at half the number of calibration samples, in this case 6, and should lead to the detection of a clear minimum in the PRESS plot (see Chap. 6).

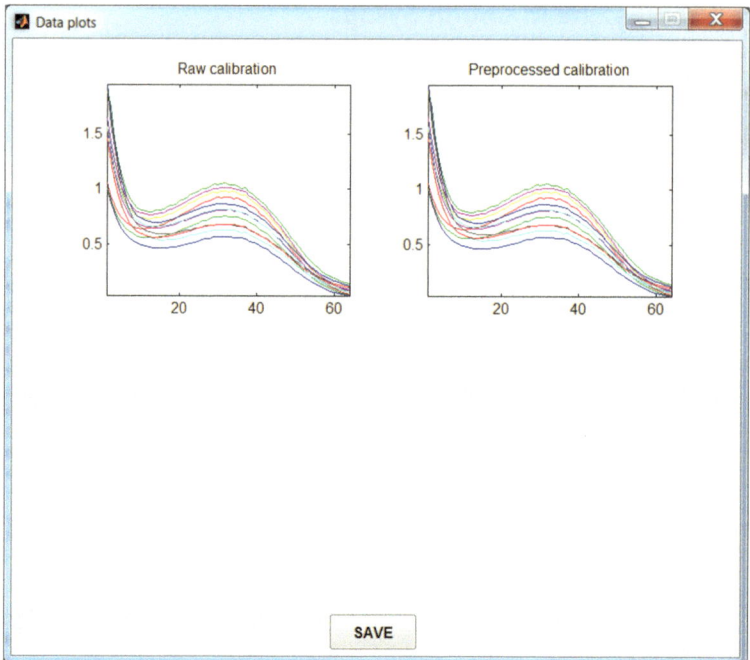

Fig. 11.7 Raw (left) and pre-processed (right) calibration spectra. In this case they are identical, because no pre-processing filter was selected (except mean centering, which is applied by default, and is not shown here). Pressing "SAVE" saves the spectra in a file in the folder "temp" of the working folder

With the screen set as in Fig. 11.9, pressing the button "CV" generates the plots of Fig. 11.10 and the table of Fig. 11.11. Figure 11.10 allows one to conclude that: (1) there is a clear minimum in the plot of PRESS values as a function of increasing number of latent variables (top plots), (2) there are no calibration outliers (the values of the F_y/F_{crit} ratios are all smaller than 1 in the left bottom plot), and (3) the predicted concentration values for the left out samples during the cross validation process are satisfactory, at least visually (right bottom plot).

On the other hand, the table of Fig. 11.11 indicates that the optimum number of latent variables to carry out this calibration is 3, in spite of the fact that the minimum PRESS in Fig. 11.10 occurs at four latent variables. This is due to the fact that the difference between the PRESS values for three and four latent variables is not significant. Finally, the RMSECV value for three latent variables, of 0.015 concentration units, is also regarded as satisfactory, in view of the fact that the average analyte concentration in the calibration samples is of ca. 2.1 units, corresponding to a relative error for the cross validation phase of only 0.7%. A list of F_y/F_{crit} ratios is also available for each calibration sample in Fig. 11.11.

Once the optimum number of latent variables (three) is estimated, we are in the position of building the PLS-1 calibration model (Fig. 11.12). Pressing the button

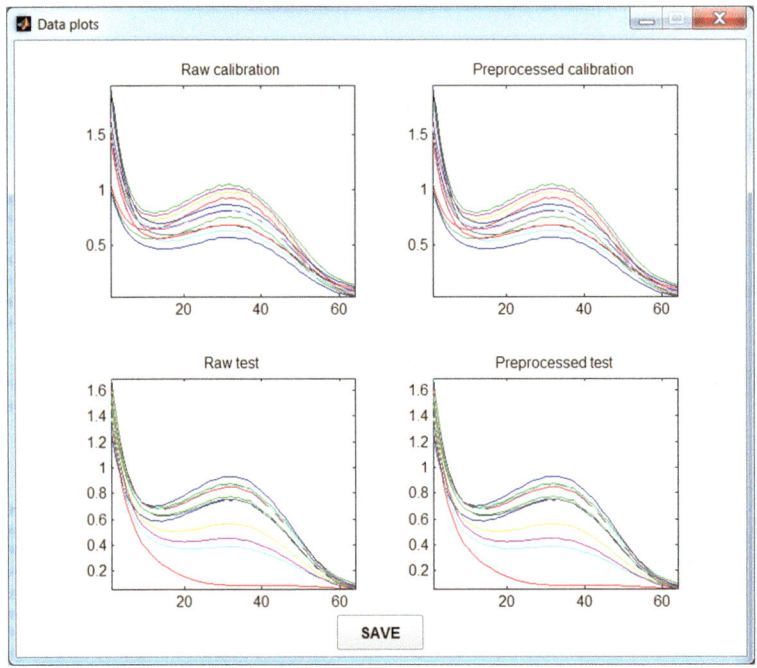

Fig. 11.8 Plot of spectra for calibration (top) and validation (or test) samples (bottom)

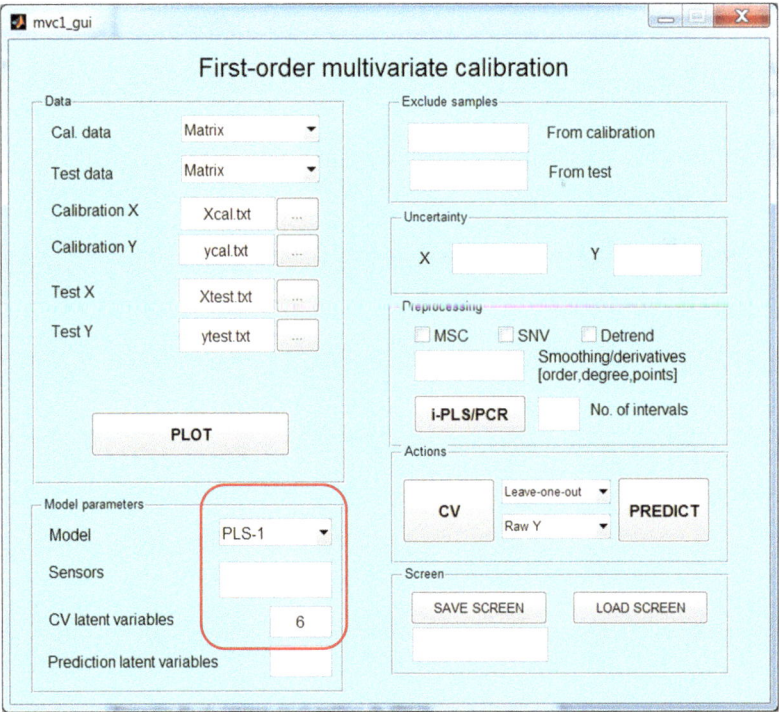

Fig. 11.9 MVC1 screen for performing LOO cross validation with the PLS-1 model. The number of maximum latent variables has been set to 6 (red box)

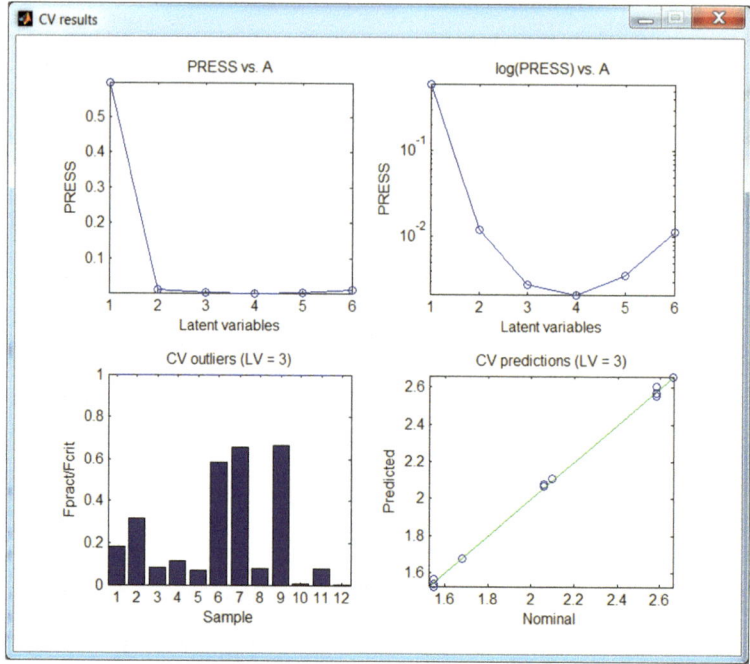

Fig. 11.10 MVC1 screen with LOO cross validation results. Top, PRESS plots as a function of the number of latent variables. Bottom, left: outlier detection. Bottom, right: cross validation prediction results (predicted vs. nominal values in the left out samples)

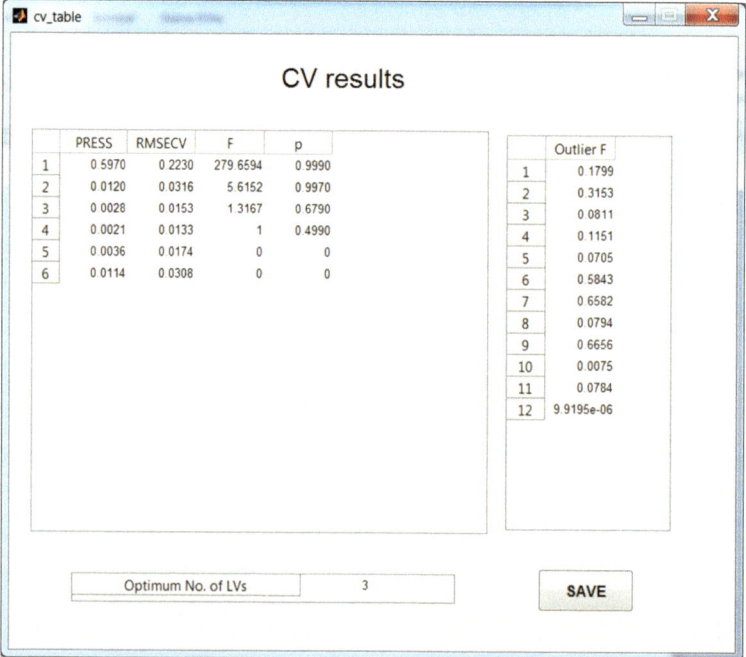

Fig. 11.11 MVC1 table with cross validation results. Pressing "SAVE" saves the results in a file in the folder "temp" of the working folder

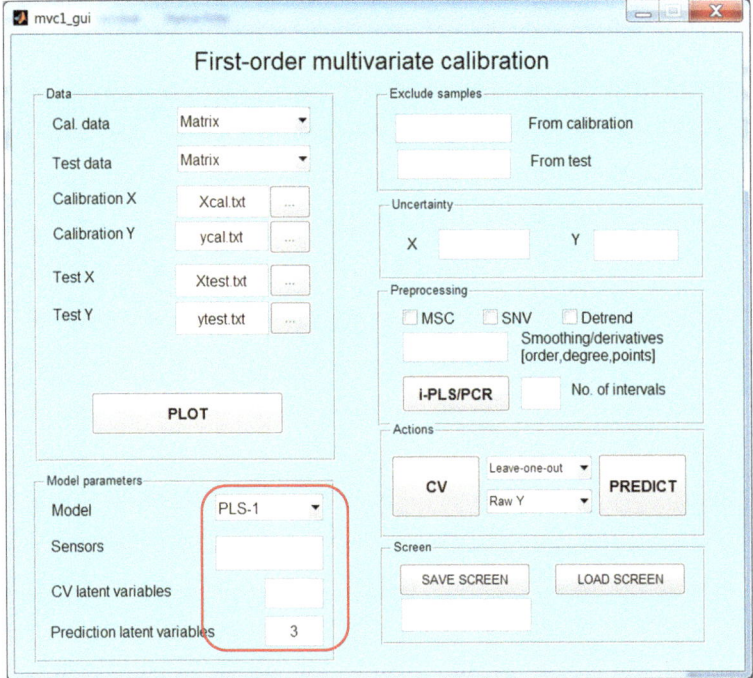

Fig. 11.12 MVC1 screen prepared to build a PLS-1 model with three latent variables (red box)

"PREDICT" provides access to the prediction phase, accompanied by abundant information on the calibration results, to be discussed below.

Figure 11.13 shows the plot of predicted concentrations vs. nominal values for the validation samples, the prediction errors (both as a function of sample index and predicted concentration), and the region of joint confidence for the slope and intercept of the plot of predicted vs. nominal values. In this latter plot, we expect the ideal point of unit slope and zero intercept to be contained within the ellipse if the analysis is accurate (González et al. 1999). Although this is not the case, in the present example the point is very close to the elliptical region, which can be considered satisfactory in view of the fact that this accuracy test is very strict.

The specific predicted values are shown in the table of Fig. 11.14, together with the theoretical prediction uncertainties and a parameter (a ratio of F values for the test samples, see Chap. 7) indicating the potential presence of outliers. This indicator is favorable for most samples, except sample No. 3, to be discussed below. Figure 11.14 also shows a table headed "Statistics," with the average absolute prediction error (RMSEP), the relative error of prediction (REP), the correlation coefficient (R^2), the calibration residues (in signal units), and the % of explained variance both in signal (X) and in concentration (y) for the calibration samples. The Durbin–Watson parameter and its associated probability (measuring the correlation among residuals, see next chapter), and the number of hidden RBF neurons, only

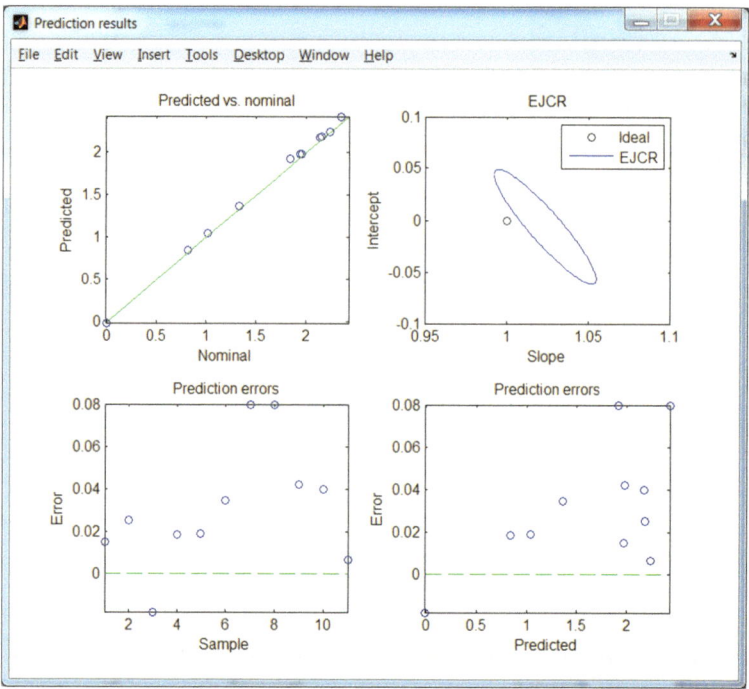

Fig. 11.13 Prediction plots for bromhexine concentration in cough syrups. Top, left: predicted vs. nominal concentrations in the validation samples. The green straight line indicates the ideal line of unit slope. Top, right: elliptical confidence joint region for the slope and intercept of the prediction plot. The black circle indicates the position of the ideal point. Bottom, left: prediction errors vs. sample number. Bottom, right: prediction errors vs. predicted concentration

applicable to calibration with neural networks, will be discussed in the next chapter. The expression "NaN" in this case is a MATLAB symbol for *not a number*.

Finally, a table with the analytical figures of merit ("AFOMs") is also provided in Fig. 11.14, as discussed in Chap. 10: sensitivity, analytical sensitivity, detection limits, and quantitation limits. It is important to notice that the estimation of the figures of merit requires one to know the uncertainty in instrumental signals and calibration concentrations. The software estimates these values from the spectral and prediction residuals of the cross validation phase, respectively. However, if the analyst has an independent experimental estimation of these uncertainties, they can be employed to compute the figures of merit. This can be indicated in the main MVC1 screen before prediction is made. Figure 11.15 shows the main screen for building a PLS-1 model, considering 0.005 units of uncertainty in instrumental signal (typical of UV-visible spectra registered in a modern spectrophotometer) and 0.01 units of uncertainty in calibration concentrations (derived from sample preparation or from the typical error for the reference technique).

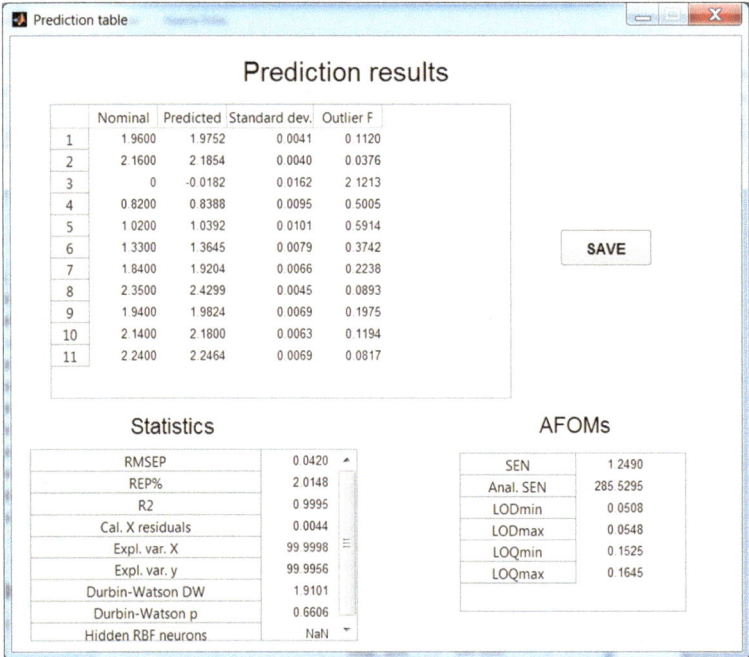

Fig. 11.14 Prediction results for bromhexine concentration in cough syrups. Top: nominal and predicted concentrations in the test samples, standard deviations, and outlier indicators. Bottom, left: prediction statistics. Bottom, right: analytical figures of merit. Pressing "SAVE" saves the results in a file in the folder "temp" of the working folder

Additional plots provided by MVC1 are shown in Fig. 11.16: the bar plot in the left top repeats the information on outliers in the test samples. The score–score plot (right top) shows the calibration and validation samples located according to the values of the second and first PLS-1 scores, which is useful for getting a quick view at the relative position of test samples in the calibration space (notice the red circle indicating the location of the test sample No. 3). Figure 11.16 also provides the spectrum of the regression coefficients (left bottom), whose analysis is favorable in the whole spectral range, in the sense that it does not present saturated or high-noise regions. Finally, the PLS-1 spectral loadings are plotted in the right bottom of Fig. 11.16, all of which display spectral-type shapes and not random noise.

All the analytical indicators point, in general terms, to the success in the calibration for this particular analyte in this type of samples. Specifically, analysts from the producing laboratory were pleased to see the relative error of prediction of ca. 2%. This is the main parameter of interest for adopting a new analytical methodology, besides other properties such as time, cost, and ease of operation.

Notice that in the present example both calibration and validation data are available. Their spectra are provided along with the corresponding nominal analyte concentrations. In the case of true unknown samples, no information on analyte

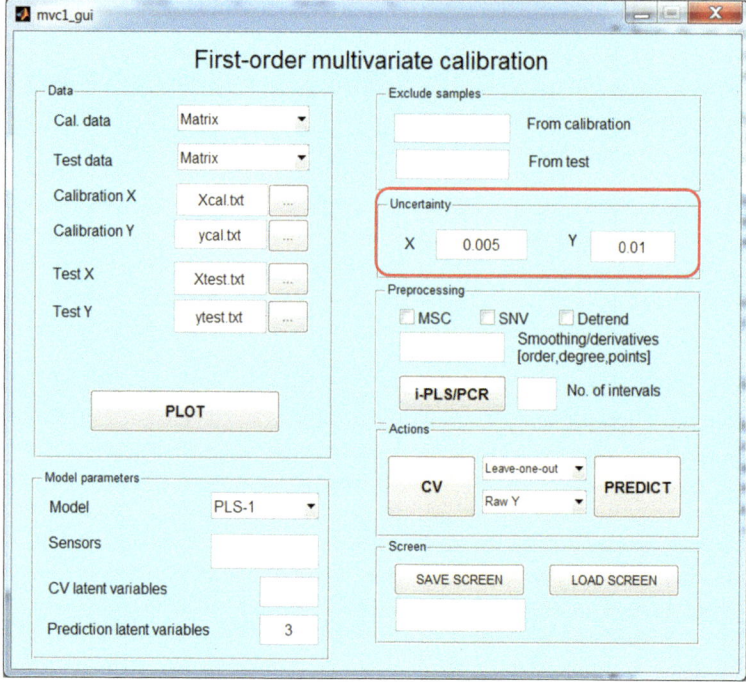

Fig. 11.15 MVC1 screen prepared to build a PLS-1 model with three latent variables, a signal uncertainty of 0.005 units and a concentration uncertainty of 0.01 units (red box)

concentration is of course available. To process this type of samples, the file containing the measured signals will be loaded from the main screen of MVC1, leaving the box for the filename with nominal concentrations empty. The prediction results will be similar to above, except those requiring nominal analyte concentrations to be estimated.

11.5   Prediction Outliers in the Bromhexine Example

We now examine in detail the case of test sample No. 3. In Fig. 11.16 (top left plot) the blue bar for this sample suggests that it is a potential outlier. Moreover, in the same figure (top right plot, red circle) the sample in question appears to be outside the calibration space, and should be considered, in principle, as not being representative of the calibration set, probably containing constituents which were not included in the calibration. All this information suggests that sample No. 3 may be a prediction outlier.

However, we know that sample No. 3 is not of this type: it is a blank sample, with a nominal analyte concentration of zero, for which the predicted concentration is quite good (ca. −0.02 units in Fig. 11.14). Why does the PLS-1 model flag it as an

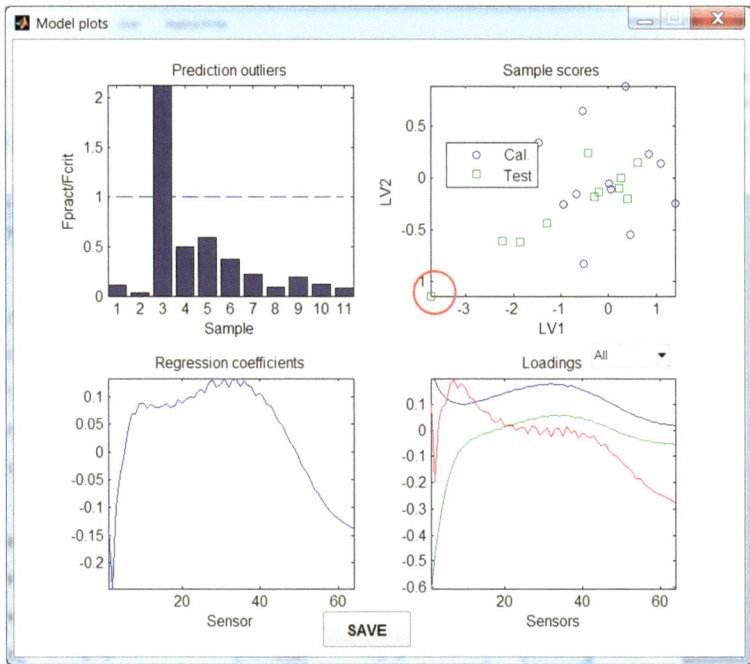

Fig. 11.16 PLS-1 prediction plots. Top, left: outlier detection. Top, right: location of the calibration and prediction samples in the score space (red circle, validation sample No. 3). Bottom, left: vector of regression coefficients. Bottom, right: the three loadings (specific loadings can be selected from the drop-down list). Pressing "SAVE" saves the results in a file in the folder "temp" of the working folder

outlier? The best answer is: because the calibration set does not contain blank samples. We appreciate here how strict is the PLS-1 model with respect to sample representativity. The calibration composition is a key factor: it should contain the analyte and the potential interferents in variable proportions, representing the composition of future samples. In the absence of a reference concentration value for this type of samples, it would not be possible to judge whether the prediction is good, or if the sample should be submitted to a different analytical methodology for prediction.

One final comment on test outliers: if they occur, either in calibration or in validation samples, it is possible to ignore these samples for calibration or prediction. For example, if the calibration sample No. 2 and the validation sample No. 3 are considered outliers, they can be discarded from the analysis as shown in Fig. 11.17. To exclude multiple samples, simply type their numerical indexes separated by blank spaces. In any case, it is always convenient to repeat the preparation of calibration outliers, especially when the calibration set has been statistically designed.

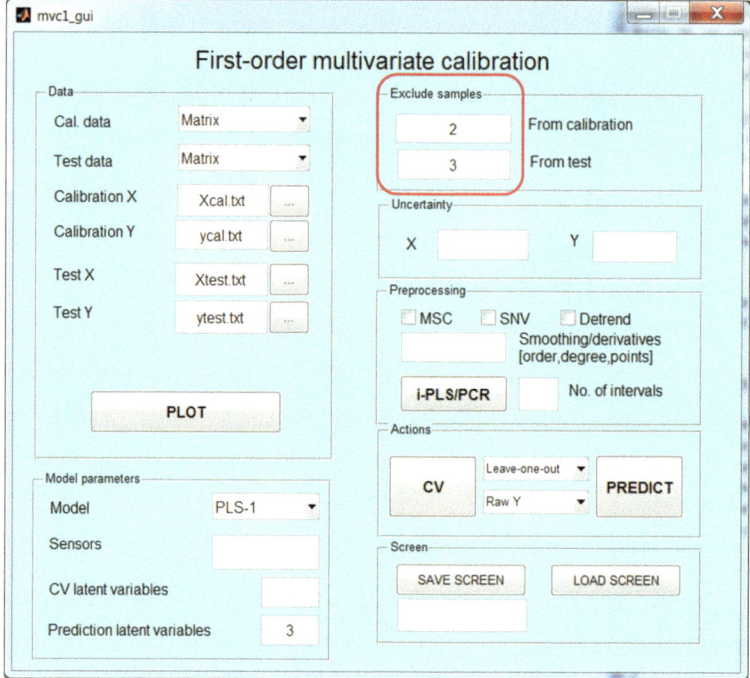

Fig. 11.17 MVC1 screen prepared to build a PLS-1 model with three latent variables, excluding the calibration sample No. 2 and the validation sample No. 3 (red box)

11.6   A Dinitro-cresol in Reaction Mixtures

In this system labeled *I5 in reactor* the analyte is 2,6-dinitro-*p*-cresol (whose commercial name is *I5*), employed in the chemical industry to stop polymerization reactions. It is added in controlled concentrations to mixtures of unsaturated and aromatic hydrocarbons, whose composition is not completely known. Both the hydrocarbons and the analyte I5 strongly absorb in the UV region, but the analyte shows a red shift which allows one to partially distinguish its signal from the background, although a considerable degree of overlapping persists (Arancibia et al. 2005).

Figure 11.18 shows the MVC1 screen corresponding to this particular example. As can be seen, the spectra were saved by the instrument in files of the type "X,Y_vectors" (reader: you can check this by opening a typical file). The calibration and validation spectra are presented in Fig. 11.19. What conclusions can be drawn a priori with respect to the working sensors for building the calibration model?

This example was specifically chosen to illustrate the advantages in selecting working spectral ranges for PLS-1 calibration. Examination of Fig. 11.19 indicates that it is not convenient to calibrate with sensors above No. 300. In the spectral

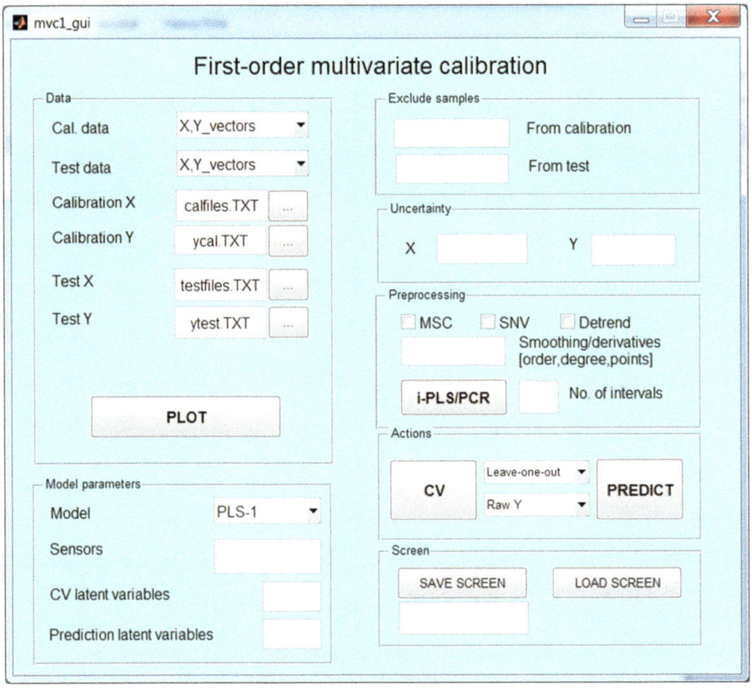

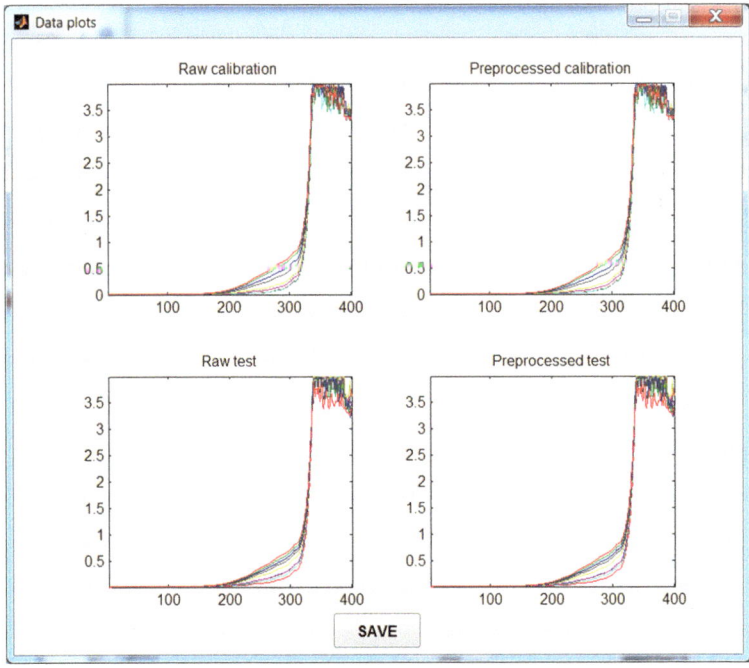

Fig. 11.18 MVC1 screen prepared to build a model for measuring the content of I5 in reaction mixtures

Fig. 11.19 Calibration and validation spectra in the system for measuring the content of I5 in reaction mixtures

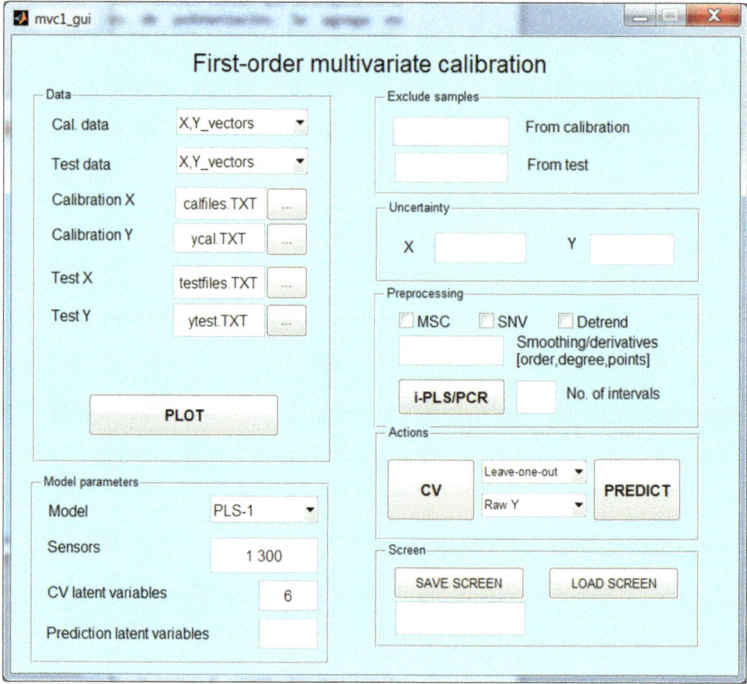

Fig. 11.20 MVC1 screen prepared for the determination of the content of I5 in the sensor range from 1 to 300. LOO cross validation will be performed with a PLS-1 model using a maximum of six latent variables

region between sensors 300 and 400 the absorbance is too high (corresponding to very short wavelengths), and the detector is saturated, generating a noise level that will be difficult to model with PLS-1. We may try to calibrate the model with the sensors from 1 to 300, for which the screen program will look as in Fig. 11.20. We will employ a maximum of six latent variables for LOO cross validation, because ten calibration samples are available. The result from this phase is that two latent variables are optimal, with reasonable analytical indicators and with no outliers. Prediction of the analyte content in the validation samples with two latent variables (Fig. 11.21) leads to good statistical indicators: the relative error is ca. 1.5%. The reader may check that the validation sample No. 6 is an outlier in prediction, and that excluding this sample from the analysis leads to a REP value of 1.1%, meaning that the error has not significantly decreased. This particular validation sample might carry a different background composition in comparison with the calibration samples, but was not further checked by the industrial laboratory. However, notice that the ratio of F values for spotting outliers in the case of the validation sample No. 6 is not significantly larger than 1, so it is likely that it is not a true outlier.

To illustrate how variable selection can be automatically performed using MVC1, and not by just visual inspection of the spectra, we will employ the i-PLS technique

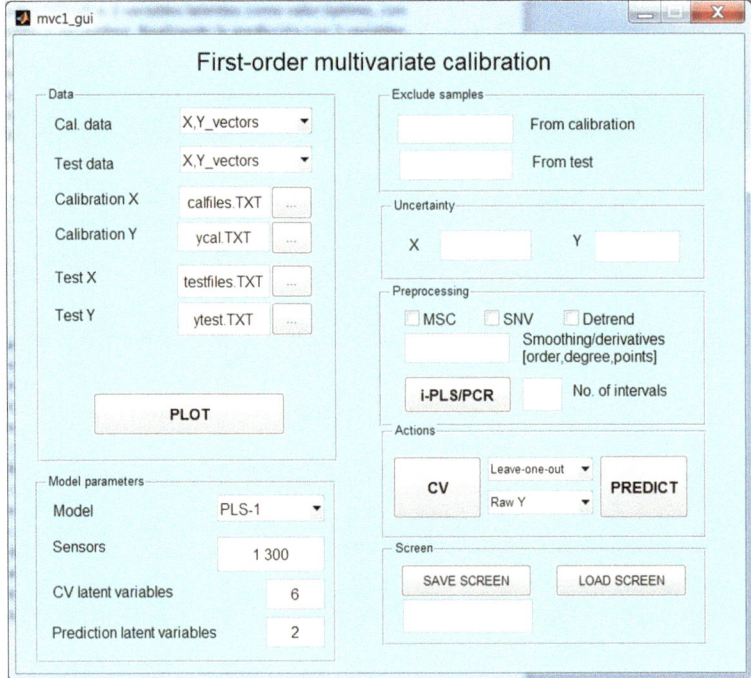

Fig. 11.21 MVC1 screen prepared for predicting the content of I5 in the validation samples, using a PLS-1 model with two latent variables in the range of sensors from 1 to 300

by means of the button "i-PLS/PCR," which allows one to build the calibration model in sequential sensor intervals. For this purpose, the number of intervals for estimating the prediction error should be provided, as shown in Fig. 11.22. Notice that a certain relationship should exist among the total number of sensors, the selected number of intervals, and the number of latent variables for building the models. In this particular case, the total number of sensors is 400, and the number of calibration samples is 10. Since i-PLS calculations are performed using Monte Carlo cross validation with 70% of the calibration samples (because full LOO cross validation is considerably slower), we are left with seven calibration samples. To be able to apply i-PLS, the number of sensors per interval should be larger than the maximum number of trial latent variables for building each model. Consequently, if the maximum number of latent variables is 4, the number of intervals should be smaller than 400/4 = 100 (rounded to the nearest integer). Otherwise, an error message is obtained (Fig. 11.23).

We may thus select 4 latent variables and 40 intervals of 10 sensors each, as in Fig. 11.22. The resulting bars (Fig. 11.24) measure the relative prediction errors in each interval (the maximum one is scaled to 1). As can be appreciated, the errors are minimal in the range of sensors from 180 to 330. The region from 1 to 180 sensors could be discarded because in this range the absorbances are negligible, and will not

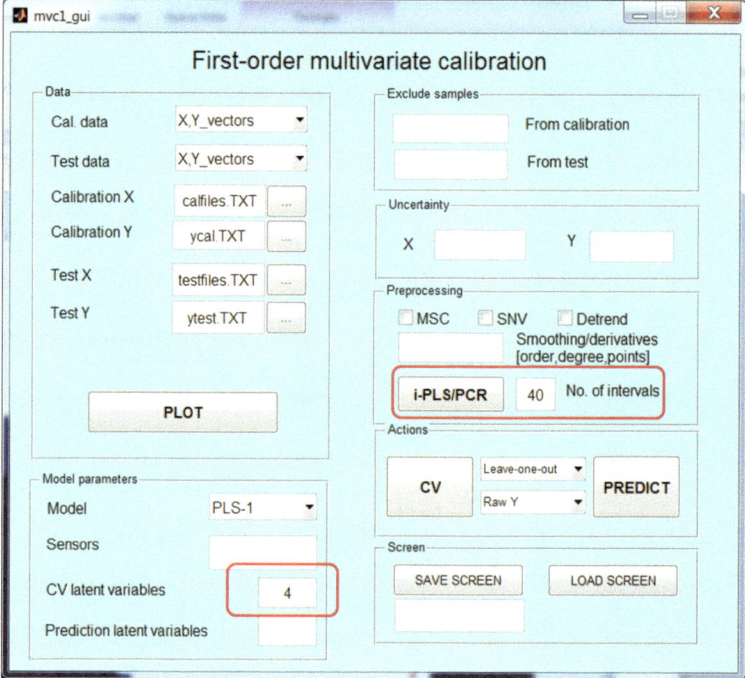

Fig. 11.22 MVC1 screen prepared to apply the variable selection method i-PLS. The red boxes show the maximum number of latent variables to be employed (4), and the selected number of intervals (40)

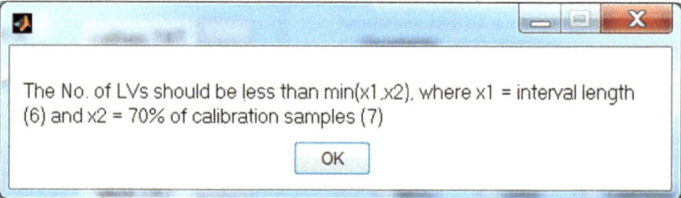

Fig. 11.23 Error message obtained by setting the number of intervals as too large to be compatible with the number of calibration samples and latent variables

significantly contribute to the model. The region above sensor 330 is problematic due to significant noise and detector saturation and should be removed from the model. With these premises, the reader may be able to build the proper model in region 180–330, using first LOO cross validation for estimating the optimum number of latent variables. Did you get better figures of merit and statistical indicators using i-PLS in comparison with the visual inspection above? Does the prediction outlier still exist?

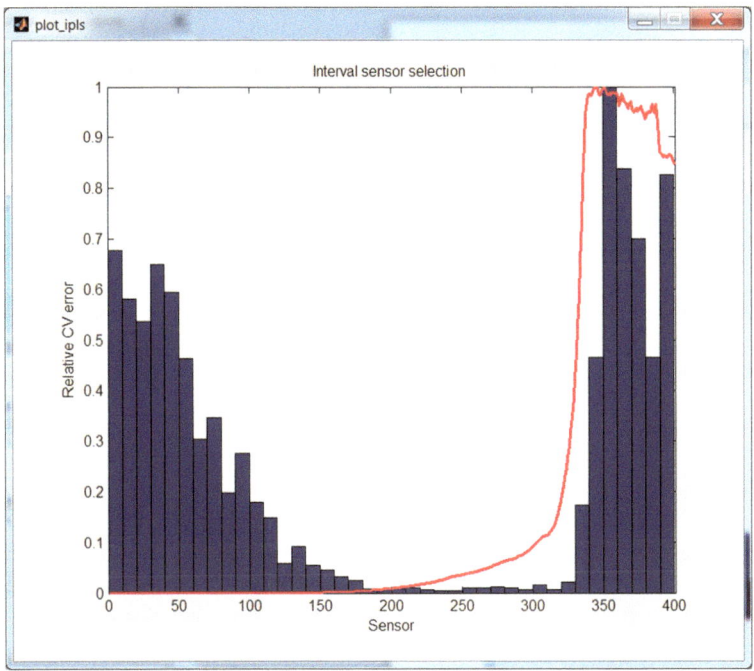

Fig. 11.24 Results after application of i-PLS. The blue bars are proportional to the prediction errors in each interval (normalized so that the maximum error is 1). The red line indicates the average calibration spectrum

11.7   Moisture, Oil, Protein, and Starch in Corn Seeds

The data for this example (labeled *Parameters in corn*) are available on the internet, as already discussed in previous chapters, and correspond to the measurement of quality parameters of corn seeds by NIR spectroscopy. We apply PLS-2 as a global model for all properties simultaneously. Since the NIR spectra were collected on the solid samples, they show a significant degree of dispersive signals. We may thus explore the different methods which are available in MVC1 for reducing dispersive effects, applying pre-processing mathematical filters.

The main MVC1 screen can be set as in Fig. 11.25. To apply the global PLS-2 model, we need a file with the calibration concentrations for the four parameters in four different columns. This file (ycal.txt) contains the nominal reference values of moisture, oil, protein, and starch, which will be our analytes No. 1, 2, 3, and 4, respectively. The first activity is cross validation. We will modify the usual procedure in two respects: (1) instead of using LOO cross validation, we will employ Monte Carlo cross validation, which is faster for large sets of calibration samples, and (2) we will not use the raw concentration values for calibration, but scaled values, in such a way that the four parameters have the same minimum and

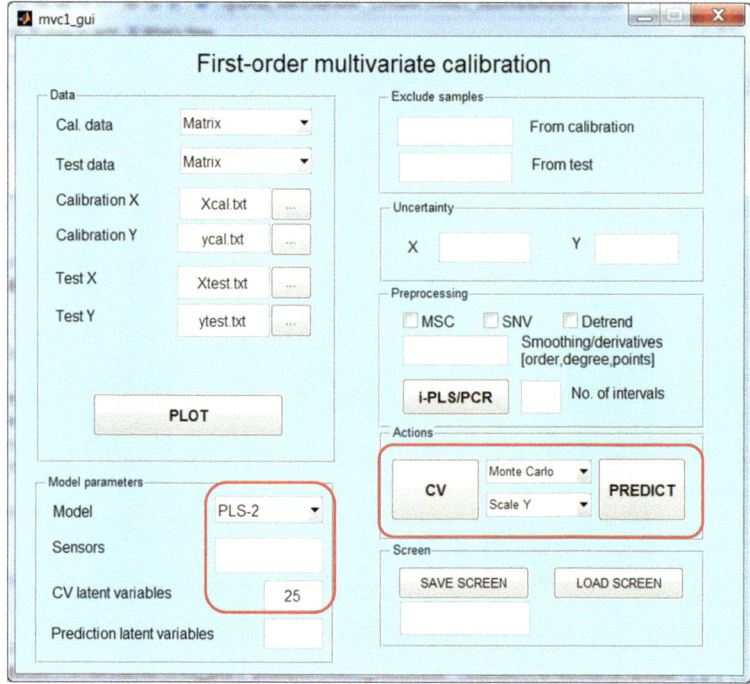

Fig. 11.25 Main screen of MVC1, prepared for carrying out Monte Carlo cross validation on NIR spectral data for measuring quality parameters of corn seeds. Scaled values of the properties to be calibrated are employed in a PLS-2 model with a maximum of 25 latent variables. The red boxes show the selection of the relevant calibration parameters

maximum values (0 and 1 in MVC1). This is done to avoid analytes or parameters having large property values dominate the PRESS over the remaining ones. In a first stage, no pre-processing filters are used, and the whole spectrum is employed for model building, since no saturation effects are noticed (reader: check this fact pressing "PLOT").

Selecting 25 latent variables as the maximum value for Monte Carlo cross validation (the number of calibration samples is 50), the optimum appears to be in the neighborhood of 19–21 latent variables (the results may vary due to the random nature of the Monte Carlo method), with an average error of ca. 0.01 units (in a scale from 0 to 1, because all parameters were scaled for this procedure). This error is reasonably small, implying only 2% of the average scaled value of 0.5 units. Reader: how many latent variables are suggested by LOO cross validation? Which is the RMSECV in this case? How does it compare with Monte Carlo cross validation in terms of relative error? Are there outliers in cross validation?

With 20 latent variables, pressing "PREDICT" provides access to the prediction plots analogous to the previous examples, except for a new drop-down window permitting the selection of the analyte of interest. For example, Fig. 11.26 shows the

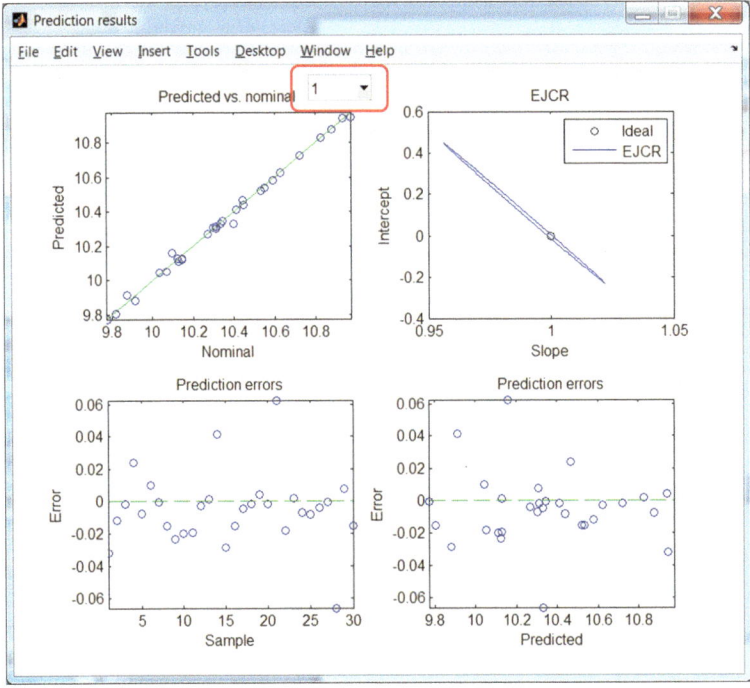

Fig. 11.26 Prediction results for analyte 1 (moisture) in corn seeds using the PLS-2 model of NIR spectra. The red box shows the drop-down list where one may select the analyte of interest

results for the analyte No. 1 (moisture); selecting another analyte index provides the corresponding information. In this table, we can also select the numerical results for each analyte: prediction, statistics, and figures of merit. In terms of relative error of prediction, we conclude that they are ca. 0.2% for moisture, 2% for oil, 1.2% for protein, and 0.3% for starch. Are these values satisfactory? Apparently yes, although this may depend on the existing protocols and regulations, and on the uncertainty associated to the reference analytical techniques used to build the multivariate models.

We now apply a mathematical pre-processing filter to decrease the effect of NIR dispersion, hoping to also decrease the number of latent variables required to build the model. A large number of combinations of pre-processing filters is possible, each with a certain optimum number of latent variables and prediction error. We only show the effect of the second derivative. Figure 11.27 shows the MVC1 main screen, prepared to perform Monte Carlo cross validation with the PLS-2 model, scaled concentration values and the second derivative for pre-processing. The latter will be estimated with a polynomial of degree 4 and a moving window of 15 sensors (Fig. 11.27); hence the three numbers "2 4 15" to be typed by the operator in the corresponding space. The pre-processed spectra can be inspected by pressing "PLOT" (Fig. 11.28).

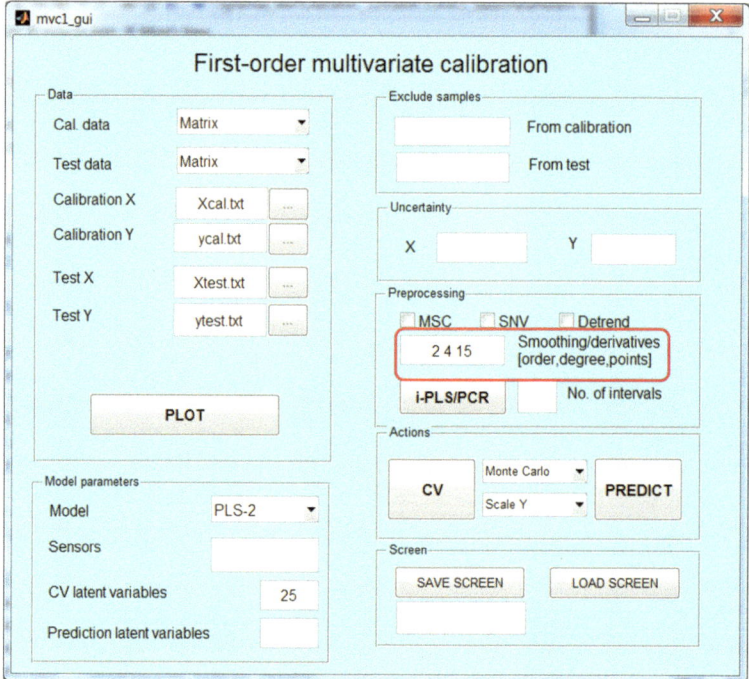

Fig. 11.27 Main screen of MVC1, prepared for carrying out Monte Carlo cross validation on NIR spectral data for measuring quality parameters of corn seeds. Scaled values of the properties to be calibrated are employed in a PLS-2 model with a maximum of 25 latent variables, and second derivative as pre-processing filter. The red box shows the Savitzky–Golay parameters (2 4 15), indicating the order of the derivative (2), the degree of the polynomial (4), and the width of the moving window (15 sensors)

The cross validation process indicates an optimum number of latent variables of 11, for which the RMSECV is 0.012 units. In comparison with the use of raw data (see above), we conclude that this model is more parsimonious, without a significant loss of predictive power, since almost the same error level is achieved with fewer latent variables. It is clear that the second derivative is effective in decreasing the impact of the dispersion on the spectral variance.

Using 11 latent variables, the PLS-2 model predicts the concentrations of all four analytes with average relative errors of ca. 1%, 2%, 2%, and 0.4% (moisture, oil, protein, and starch, respectively). The predictive ability is maintained, except for moisture, although a relative error of 1% is still acceptable.

There is also the possibility of applying separate PLS-1 models for each of the four properties, as already commented (Table 5.1 of Chap. 5 using PCR with raw data, Table 7.2 of Chap. 7, comparing PCR and PLS-1 with raw data, and Table 9.2 of Chap. 9 using PLS-1 for total oil content using second derivative pre-processing). Table 11.2 summarizes the results for the oil content. During a night of insomnia, the reader could build a similar table for the remaining three properties, selecting

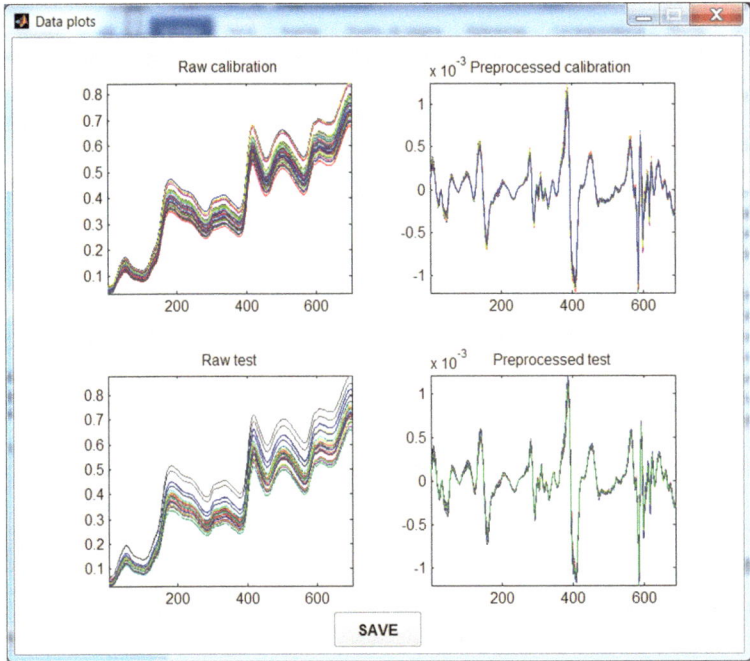

Fig. 11.28 Raw and second derivative NIR spectra of the set of corn seeds

Table 11.2 Relative errors of prediction (REP) for the determination of oil content in corn seeds using various multivariate models

Model	Math filter	Number of latent variables	REP (%)
PLS-1	None	21	1.1
	Second derivative	11	1.2
PLS-2	None	20	2.0
	Second derivative	11	2.0

optimal working regions with i-PLS, the best pre-processing filter and number of latent variables.

What conclusions can be drawn from Table 11.2? Simple: PLS-1 is to be preferred, and the second derivative filter produces a more parsimonious model without significant loss of predictive power.

11.8 Additional MVC1 Models

In the drop-down list of models, the reader will find additional possibilities beyond PLS-1 and PLS-2. One of them is PCR, already discussed in Chap. 5. Another available model is MLPCR (maximum likelihood PCR), which takes into account

Table 11.3 Commercial software for first-order multivariate calibration

Program	Vendor	Web site
The Unscrambler	Camo	www.camo.com
Pirouette	Infometrix	www.infometrix.com
PLS_Toolbox	Eigenvector	www.eigenvector.com
Statistics and machine learning toolbox[a]	The Mathworks	www.mathworks.com

[a]Requires MATLAB to be installed

the structure of the instrumental noise, and is recommended for cases where the uncertainty in signals is not of the iid type. We will not discuss the use of this model in this book; the reader is directed to the specialized literature on the subject (Schreyer et al. 2002).

Finally, the RBF model corresponds to a type of neural network (radial basis functions), developed for systems whose behavior is non-linear with respect to concentrations or target properties. We will discuss the RBF model in the framework of MVC1 in the next chapter.

11.9   Other Programs

Many computer programs exist, both free and commercial, to carry out first-order multivariate calibration, of the type described in this book. Table 11.3 collects some commercial software, the vendors, and the web sites where to find them.

It is also interesting to mention the cell phone application MVC written in Android, which allows one to load data, build PCR and PLS models, and plot prediction results (Parastar and Shaye 2015). It is freely available in http://sharif. edu/~h.parastar/software.html.

11.10   Exercises

1. Process the data set named *Tetracycline in serum*, corresponding to the determination of the antibiotic tetracycline in human sera by means of synchronous fluorescence spectra (Goicoechea and Olivieri 1999b). In the folder where the data are provided, the reader will find a help text file, indicating the nomenclature of filenames, the data type, and additional useful information on this system.

   We suggest the following working protocol:

   (a) Visually inspect the spectra ("PLOT"). Are there saturated or high-noise spectral regions which would merit variable selection? Try first discarding them from the MVC1 screen, and then apply i-PLS/PCR. How do the two results compare?

   (b) Carry out LOO or Monte Carlo cross validation, depending on the number of calibration samples. Scale the concentrations if you are using the PLS-2 model, otherwise use raw concentration data.

(c) After cross validation, check for the presence of outliers in calibration. If they indeed occur, and the F ratios are significant (i.e., larger than 3), exclude the problematic samples from the calibration set from the MVC1 screen and re-run cross validation. Do the results improve? If they do not significantly improve, keep the samples in the calibration set.

(d) Set the optimum number of latent variables for building the calibration model.

(e) Predict the analyte concentrations in the validation set and study the statistical indicators and figures of merit.

(f) Check for the presence of outliers in the validation samples. Follow the same rules as for calibration outliers.

(g) Predict the analyte concentrations in true unknown samples, if they are available.

2. Process the data set named *Octane in gasolines*, corresponding to the determination of the octane number of gasoline samples using NIR spectroscopic data (Boschetti and Olivieri 2004). Follow the same indications as in the previous exercise.

References

Arancibia, J.A., Martínez Delfa, G., Boschetti, C.E., Escandar, G.M., Olivieri, A.C.: Application of partial least-squares spectrophotometric multivariate calibration to the determination of 2-sec-butyl-4,6-dinitrophenol (dinoseb) and 2,6-dinitro-p-cresol in industrial and water samples containing hydrocarbons. Anal. Chim. Acta. **553**, 141–147 (2005)

Boschetti, C.E., Olivieri, A.C.: A new genetic algorithm applied to the near-infrared analysis of gasolines. J. Near Infrared Spectrosc. **12**, 85–91 (2004)

Goicoechea, H.C., Olivieri, A.C.: Determination of bromhexine in cough-cold syrups by absorption spectrophotometry and multivariate calibration using partial least-squares and hybrid linear analyses. Application of a novel method of wavelength selection. Talanta. **49**, 793–800 (1999a)

Goicoechea, H.C., Olivieri, A.C.: Enhanced synchronous spectrofluorometric determination of tetracycline in blood serum by chemometric analysis. Comparison of partial least-squares and hybrid linear analysis calibrations. Anal. Chem. **71**(19), 4361–4368 (1999b)

Goicoechea, H.C., Olivieri, A.C.: MULTIVAR. A program for multivariate calibration incorporating net analyte signal calculations. Trends Anal. Chem. **19**, 599–605 (2000)

González, A.G., Herrador, M.A., Asuero, A.G.: Intra-laboratory testing of method accuracy from recovery assays. Talanta. **48**, 729–736 (1999)

Olivieri, A.C., Goicoechea, H.C., Iñón, F.A.: MVC1: an integrated Matlab toolbox for first-order multivariate calibration. Chemom. Intell. Lab. Syst. **73**, 189–197 (2004)

Parastar, H., Shaye, H.: MVC app: a smartphone application for performing chemometric methods. Chemom. Intell. Lab. Syst. **147**, 105–110 (2015)

Schreyer, S.K., Bidinosti, M., Wentzell, P.D.: Application of maximum likelihood principal components regression to fluorescence emission spectra. Appl. Spectrosc. **56**, 789–796 (2002)

Abstract

An introduction to calibration of non-linear systems with artificial neural networks is provided. Detailed information is given on calibration using radial basis functions. The latter are interpreted based on the concept of data linearization by projection onto a non-linear space.

12.1 Linear and Non-linear Problems

In previous chapters, the focus has been mainly directed to linear systems, where the relationship between multivariate signal and analyte concentration is linear. The golden rule for multivariate models comes from an old parsimony principle: *linear models for linear systems, non-linear models for non-linear systems*. Linear models are simpler, are based on well-known physicochemical laws, are reliable, and show well-defined figures of merit. They should be preferred when the system is linear.

One sensible course of action if non-linearity is suspected may be the following: start by applying a linear model (PCR, PLS-1, PLS-2) to the problem at hand. Why? Because most spectral signals vary in a linear fashion with respect to analyte concentrations. If the model results are satisfactory, and they do not suggest the presence of non-linearity (see next section), it would not make much sense to move to non-linear models. Conversely, if the analysis of the linear model indicators suggests the presence of non-linearity, it would not be adequate to apply linear models. A non-linear approach should instead be taken to process the data. There are many such models available, one of which will be discussed in this chapter.

© Springer Nature Switzerland AG 2018

A. C. Olivieri, *Introduction to Multivariate Calibration*,

https://doi.org/10.1007/978-3-319-97097-4_12

12.2 Multivariate Non-linearity Tests

Different tests have been discussed in the literature for studying the presence of non-linearities in the multivariate signal–concentration relationship (Centner et al. 1998). A simple alternative is to first build a PLS-1 model with the optimum number of latent variables, predicting the concentration of the analyte of interest in a set of validation samples. If the system is not linear, the prediction residuals will show significant correlations among them, when ordered according to increasing predicted concentration values. To avoid false visual impressions, the existence of correlations can be detected by the Durbin–Watson statistical test (Durbin and Watson 1950), which consists in estimating the following DW indicator:

$$\mathrm{DW} = \frac{\sum\limits_{n=1}^{N\mathrm{val}-1} \left(r_{n+1} - r_n\right)^2}{\sum\limits_{n=1}^{N} r_n^2} \tag{12.1}$$

where r_n is the nth prediction residue. The DW indicator will be high in the case of uncorrelated residuals, because a large number of differences between positive and negative values will occur. On the other hand, for correlated residuals DW will be small, because series of positive and negative values will occur, with a relatively small number of differences between successive residues. How can these DW values be statistically analyzed? The Durbin–Watson indicator DW has an associated probability p: if $p < 0.05$ the null hypothesis is rejected, indicating correlations among residuals, and vice versa.

12.3 A Durbin–Watson Algorithm

A very short MATLAB code allows one to estimate DW and the associated probability, as shown in Box 12.1. Notice that one should first sort the prediction values in increasing order, and then implement Eq. (12.1) by applying a built-in MATLAB routine, which gives the DW value and its associated probability.

Box 12.1

A Durbin–Watson algorithm. The input variables are "ynom," the vector of nominal analyte concentrations and "ypred," the one of predicted concentrations.

```
[orderedres,indexes]=sort(ypred);
res=ypred-ynom;
[p,dw]=dwtest(ypred(indexes)-ynom(indexes),ypred(indexes));
```

12.4   Non-linear Relationships and Projections

We now study a useful method to cope with non-linear systems. A simple univariate example will be analyzed before moving to the multivariate world. Suppose we want to fit the values of y of Fig. 12.1a, which vary in a non-linear fashion with respect to x. How can the x–y relationship be mapped if we do not know the exact mathematical expression connecting the variables? What we need is a universal approximation to non-linear relationships. Fortunately this approach exists, and is based on the concept of *projection* of the data onto a non-linear space, fitting the values of y to a linear combination of Gaussian functions of x:

$$y = \sum_{g=1}^{G} w_g \exp\left[-\frac{(x - c_g)^2}{\sigma_g^2}\right] \tag{12.2}$$

where G is the number of Gaussian functions, w_g are the coefficients of the linear combination, and c_g and σ_g the Gaussian centers and widths. A suitable set of w_g, c_g, and σ_g values should be able to do the job. Since the relationship between y and the Gaussian functions in Eq. (12.2) is linear, we may interpret Eq. (12.2) as saying that the set of x values has been first projected onto a G-dimensional non-linear Gaussian space to linearize the problem.

For the sake of illustration, ten different values of the independent variable x were selected (green circles in Fig. 12.1a) to adjust the parameters in Eq. (12.2) using only three Gaussian functions (all σ_g values were considered identical for simplicity). With the fitted parameters, the values of y can be successfully predicted, as shown in Fig. 12.1b, nicely demonstrating the ability of Eq. (12.2) to reproduce the basic non-linear relationship between the two variables.

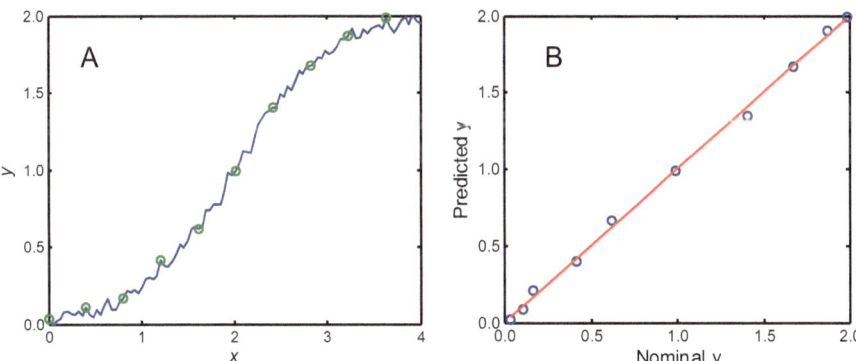

Fig. 12.1 (**a**) A non-linear function y including random noise (blue line), mapped at ten selected values of x (green circles). (**b**) Predicted vs. nominal values of y (blue circles) using the non-linear projection approach onto three Gaussian functions. The expression is $y = 0.01 \exp\left(-\frac{x^2}{4}\right) - 0.11 \exp\left[-\frac{(x-1.61)^2}{4}\right] + 2.01 \exp\left[-\frac{(x-3.64)^2}{4}\right]$. The red line indicates the perfect fit

We now move to multivariate calibration. Recall our discussion in Chap. 6 on the use of polynomials of increasing degree to map a non-linear relationship between two variables. One way of coping with mild multivariate non-linearities is to add a quadratic term to the PLS predictive equation, moving from the linear PLS model (Chap. 7):

$$y = \mathbf{v}^{\mathrm{T}} \mathbf{t}_A = \sum_{a=1}^{A} v_a t_a \tag{12.3}$$

to a PLS model with additional quadratic terms:

$$y = \sum_{a=1}^{A} \left({}^{1}v_a t_a + {}^{2}v_a t_a^2 \right) \tag{12.4}$$

where ${}^{1}v_a$ and ${}^{2}v_a$ are the regression coefficients for the quadratic approximation, estimated during the calibration phase, in a similar manner to the v_a coefficients of Eq. (12.3). Notice that we have removed the subscript n, corresponding to the specific nth analyte of interest to avoid confusion.

The model defined by Eq. (12.4) is known as quadratic PLS (qPLS) (Baffin et al. 1999), and is a good choice when mild deviations from linearity exist. Why? Because slight deviations of the strict linearity can be modeled by a quadratic term. However, for general non-linear relationships (most probably of unknown nature), there is no guarantee that Eq. (12.4) will be able to do the job.

Let us assume, for simplicity, that the number of latent variables capturing the spectral variance by a suitable PCA study of a given calibration data matrix is 2 ($A = 2$). How can the two PCA scores t_1 and t_2 for a given sample be projected onto a Gaussian space? Assuming that three Gaussian functions are needed, the vector of projected scores can be expressed as:

$$\mathbf{d} = \begin{bmatrix} \exp\left[-\dfrac{(t_1 - c_{11})^2}{\sigma^2}\right] \exp\left[-\dfrac{(t_2 - c_{12})^2}{\sigma^2}\right] \\[2ex] \exp\left[-\dfrac{(t_1 - c_{21})^2}{\sigma^2}\right] \exp\left[-\dfrac{(t_2 - c_{22})^2}{\sigma^2}\right] \\[2ex] \exp\left[-\dfrac{(t_1 - c_{31})^2}{\sigma^2}\right] \exp\left[-\dfrac{(t_2 - c_{32})^2}{\sigma^2}\right] \end{bmatrix}$$

$$= \begin{bmatrix} \exp\left\{-\left[\dfrac{(t_1 - c_{11})^2}{\sigma^2} + \dfrac{(t_2 - c_{12})^2}{\sigma^2}\right]\right\} \\[2ex] \exp\left\{-\left[\dfrac{(t_1 - c_{21})^2}{\sigma^2} + \dfrac{(t_2 - c_{22})^2}{\sigma^2}\right]\right\} \\[2ex] \exp\left\{-\left[\dfrac{(t_1 - c_{31})^2}{\sigma^2} + \dfrac{(t_2 - c_{32})^2}{\sigma^2}\right]\right\} \end{bmatrix} \tag{12.5}$$

where **d** is the vector resulting from the projection (size 3×1), and each element of **d** is a Gaussian function, whose argument is a combination of the scores t_1 and t_2. The Gaussian functions are centered at different values, depending on the score and on the specific element of **d**: there are six center values c_{ga} (where g defines the element of **d** and a the score), and a single value of the Gaussian width (σ).

Why three Gaussian functions? Because this number was found to be optimal for the specific problem at hand. Below we will indicate how to estimate the dimensionality of the Gaussian space by statistical techniques in a general case. After linearizing the system by projection, the vector **d** can be employed to predict the analyte concentration by the following linear expression:

$$y = \mathbf{w}^T \mathbf{d} \tag{12.6}$$

where **w** is a vector of regression coefficients (size 3×1), whose elements can be estimated by least-squares. Equation (12.6) is analogous to the prediction PLS expression, with the vector of regression coefficients **v** replaced by **w**, and the score vector **t** for a given sample replaced by its projection onto a Gaussian space **d**.

In a general case where the number of latent variables is A and the dimensionality of the Gaussian space G, the individual d_g elements of **d** are given by:

$$d_g = \exp\left[-\sum_{a=1}^{A} \frac{(t_a - c_{ga})^2}{\sigma^2}\right] \tag{12.7}$$

The centers of the Gaussian functions are contained in a matrix **C** of size $G \times A$, whose generic element is represented by c_{ga}. In principle, an additional set of $G \times A$ values of Gaussian widths will also be required, but in the usual formulation the widths of the Gaussian functions are all identical, so that a single value σ is needed, as in Eq. (12.7). In Eq. (12.6), **w** and **d** are both vectors of size $G \times 1$; the former is estimated during the calibration phase, the latter corresponds to a test sample. Combining Eqs. (12.6) and (12.7), the prediction phase is condensed into the following expression:

$$y = \sum_{g=1}^{G} w_g \exp\left[-\sum_{a=1}^{A} \frac{(t_a - c_{ga})^2}{\sigma^2}\right] \tag{12.8}$$

To carry the analogy with PLS calibration further, assume that the dimensionality of the Gaussian space G (the number of Gaussian functions), their centers and widths are all known. The weights can then be estimated from a set of calibration samples of known analyte concentrations, which are collected in the usual calibration vector **y**. The raw matrix of calibration spectra **X**, in turn, is first compressed and truncated to the calibration score matrix $\mathbf{T}_A$, and the latter projected onto the Gaussian space to give a compressed-truncated-and-projected matrix **D** (size $G \times I$) before applying an inverse least-squares regression expression. Each column of **D** has the form of Eq. (12.5) for each calibration sample. Some researchers call **D** the design matrix, a term which is analogous to the calibration matrix in PLS; in fact, the design matrix is the result of compressing, truncating, and projecting the calibration matrix.

The corresponding expression for the calibration phase is therefore:

$$\mathbf{y} = \mathbf{D}^{\mathrm{T}}\mathbf{w} + \mathbf{e} \tag{12.9}$$

from which $\mathbf{w}$ can be estimated, as in the ILS model of Chap. 3, by least-squares:

$$\mathbf{w} = \left(\mathbf{D}\mathbf{D}^{\mathrm{T}}\right)^{-1}\mathbf{D}\mathbf{y} \tag{12.10}$$

The requirements for inverting the $G \times G$ square matrix $\left(\mathbf{D}\,\mathbf{D}^{\mathrm{T}}\right)$ in Eq. (12.10) are, as expected: (1) the number of calibration samples (I) should be larger than G, the dimensionality of the Gaussian space and (2) the columns of $\mathbf{D}$ should not show significant correlations (reader: why?). To avoid problems with matrix inversion, a clever resource is to use ridge regression (see Chap. 3):

$$\mathbf{w} = \left(\mathbf{D}\mathbf{D}^{\mathrm{T}} + \lambda\mathbf{I}\right)^{-1}\mathbf{D}\mathbf{y} \tag{12.11}$$

where $\mathbf{I}$ is an identity matrix of proper size and λ is a small number, which is also adjusted during the training phase. Once the $\mathbf{w}$ vector is estimated, prediction for new samples proceeds as in Eq. (12.8).

Equation (12.8) is highly useful for coping with non-linear systems, and is an attractive alternative when PLS or qPLS is not adequate. If the reader had never heard about artificial neural networks (ANN), the theoretical discussion of non-linear modeling would end here, and we would smoothly move to describe non-linear data sets. However, a brief introduction to neural networks is necessary, mainly because the subject has captured the public imagination. The reason is easy to grasp: neural networks are advertised to be able to mimic the behavior of the human brain.

12.5   Artificial Neural Networks

The acronym ANN describes a set of algorithms capable of modeling non-linear relationships with high efficiency. The terminology employed in the framework of these models surprises the reader with similarities, sometimes exaggerated, with the way in which the human brain works: neurons, inter-neuron connections, neuron activation, artificial intelligence, network learning, etc. There might be some link between the way in which ANN are presented to the public and what is known about the brain, but if the neural terminology is replaced by a more technical and less flowery prose, the connection may be easily lost. In the analytical multivariate calibration context, ANN are useful tools employed to model non-linear relationships between multivariate signals and analyte concentrations or sample properties as targets. In this context, they can be viewed as nothing else than calibration models with a number of adjustable parameters, which are able to universally fit non-linear relations.

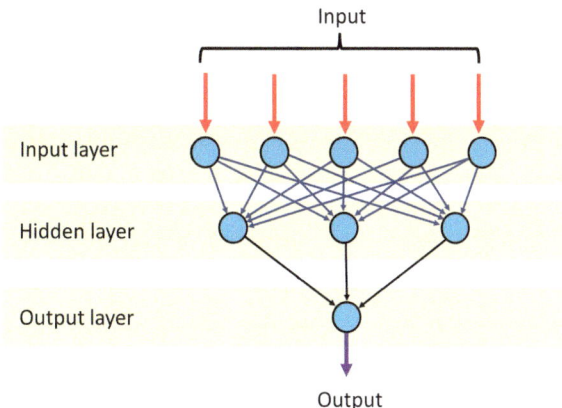

Fig. 12.2 A typical neural network architecture with three layers of neurons

The key operating unit of an ANN is the neuron. Independently of its biological meaning, in multivariate calibration problems a neuron is a function receiving an input and generating an output through a suitable mathematical expression. In artificial intelligence terms, one would say that the neuron is *activated* by the input, and will *transmit* the output to another neuron.

In the usual jargon, ANN is said to have an architecture composed of neuron layers: the input, the hidden, and the output layers (Fig. 12.2). In the training or learning ANN phase, the input layer receives the values of the instrumental signals for the training (calibration) samples, which may be the raw data or the truncated score matrix. The input process is represented by the red arrows in Fig. 12.2: each neuron is activated with the signal value for each sample, i.e., the signal is the argument of a non-linear transfer function (usually a Gaussian or a hyperbolic tangent). The function value is the output of the input layer, which is transmitted to the hidden layer (blue arrows in Fig. 12.2):

$$\text{Output from an input neuron} = f(\text{input value}) \qquad (12.12)$$

where f represents the non-linear transfer function.

The neurons of the hidden layer are then activated with a weighted average of the input values, which is the argument of the transfer function:

$$\text{Output from a hidden neuron} = f(\text{weighted average of the input values})$$

$$(12.13)$$

The weights in Eq. (12.13) are adjustable parameters, to be estimated during the training phase of the network. The outputs from the hidden layer are then transmitted to the final output layer (black arrows in Fig. 12.2), linearly combined with additional weights, and used as argument of the transfer function:

$$\text{Output from the output neuron} = f(\text{weighted average of the input values})$$

$$(12.14)$$

The weights of this final average are also adjustable parameters. Finally, the output neuron yields the analyte concentration in the sample (violet arrow in Fig. 12.2).

The network learns the relation between signals and concentration by adjusting all the weights for the inter-neural connections, in such a way that the final network output is close to the nominal concentration value for the analyte. There are various methods to optimize the network parameters, which are described in the literature on ANN, either as a general topic (Zupan and Gasteiger 1999; Haykin 1999), or as applied to analytical chemistry in particular (Ni et al. 2014; Despagne and Massart 1998).

12.6 Radial Basis Functions

A useful type of artificial neural networks having three layers as in Fig. 12.2 is known as radial basis functions (RBF). The operations taking place in this network are simpler than the above description, as we shall see.

As in all ANN architectures, the input layer accommodates the significant scores of the matrix of instrumental data for the calibration samples, and transmits them directly to the hidden layer. The transfer function activating the hidden neurons is a Gaussian function, characterized by the values of the centers and widths. Sound familiar? If the net is composed of A input neurons, each of them receives the ath score for a given sample (t_a). On the other hand, if there are G hidden neurons, then $G \times A$ Gaussian functions will be required to activate the hidden layer. The centers of these functions are contained in a matrix $\mathbf{C}$ of size $G \times A$, whose generic element is represented by c_{ga}. In principle, an additional set of $G \times A$ values of Gaussian widths will also be required, but in the usual formulation of RBF networks, the widths of the Gaussian functions are all identical, so that a single value σ is needed. The reader might have noticed, at this point, that the number of hidden neurons is equal to the dimensionality of the Gaussian space discussed in Sect. 12.4.

The final activating function for the RBF output neuron is linear, thus the network output is a linear combination of the values delivered by the hidden neurons. If the weights of this linear combination are called w_g, a relatively simple equation can be produced for the network output y as a function of the input values and the network parameters:

$$y = \sum_{g=1}^{G} w_g \exp\left[-\sum_{a=1}^{A} \frac{\left(t_a - c_{ga}\right)^2}{\sigma^2} \right] \tag{12.15}$$

Equation (12.15) is identical to Eq. (12.8). We may thus see Eq. (12.15) as the result of a projection of the sample data (the scores) onto a non-linear Gaussian space, followed by a linear combination using appropriate regression coefficients. In this alternative vision of the RBF network, there are no neurons, connections, or activating functions, but a projection of non-linear data onto a non-linear space, with the aim of linearizing the problem. This latter interpretation bears no relationship with the brain-related view of artificial neural networks. However, in the remainder of this chapter we will continue using the network nomenclature, for consistency with the literature (and because it sounds fancier and more poetic than a projection onto a non-linear space).

Beyond these rather philosophical considerations, from the operational perspective it is necessary to estimate various network parameters: (1) the number of hidden neurons (G), (2) the centers ($\mathbf{C}$) and widths (σ) of the Gaussian functions, and (3) the weights ($\mathbf{w}$) of the final linear combination. This optimization can be carried out sequentially, first fixing the number of hidden neurons, widths and centers, and then finding the weights. There are several alternative procedures for achieving these goals, one of which is explained in the next section.

12.7 An RBF Algorithm

A MATLAB algorithm for RBF calibration and prediction is given in Box. 12.2. We hope to demystify the common view that neural networks are too complex to be included within the framework of an introductory text to multivariate calibration. Admittedly, some ingredients are missing in Box 12.2: how to select the number of hidden neurons, and the centers and widths of the Gaussian functions. The idea is to show that the core RBF computations are not that complex as one may first guess.

Box 12.2
The following codes allow one to perform calibration and prediction with RBF. It is assumed that the calibration data matrix has been previously subjected to PCA to find the scores. The input variables are "T," the matrix of calibration scores (scaled so that the absolute minimum and maximum are 0 and 1, respectively); "yn," the vector of calibration concentrations (also scaled between 0 and 1); and "t," the vector of test sample scores (scaled according to the numerical scale used for "T"). Additionally, the number of hidden neurons or Gaussian dimensionality ("G"), the matrix of centers ("C"), and the Gaussian widths ("s") are also required to be known. The output is "y," the predicted analyte concentration.

```
% Calibration
for i=1:size(T,1)
    for g=1:G
        Z(i,g)=0;
        for a=1:A
            Z(i,g)=Z(i,g)+(T(i,a)-C(g,a))^2/s^2;
        end
        D(i,g)=exp(-Z(i,g));
    end
end
```

(continued)

Box 12.2 (continued)

```
        end
        w=(D'*D+1e-9*eye(G))\D'*yn;
        % Prediction
        for g=1:G
            Zt(g)=0;
            for a=1:A
                Zt(g)=Zt(g)+(t(a)-C(g,a))^2/s^2;
            end
        end
        y=exp(-Zt)*w;
```

12.8 RBF Networks in MVC1

In the MVC1 software, the RBF neural network has been implemented in a similar manner to that described by Orr (1996). The estimation of the network parameters is equivalent to training the net, which requires a set of calibration spectra and the nominal values of the analyte concentrations or sample properties. In this sense, the activity is analogous to that performed during calibration with the linear counterparts PCR or PLS. The objective of the training phase is to have a set of network parameters minimizing the average calibration error, with a minimum of hidden neurons (to avoid over-training of the network, equivalent to over-fitting in PLS calibration). In any case, recall that the number of hidden neurons (G) should be smaller than the number of calibration samples (I).

The first phase of RBF calibration is to set the number of PCA latent variables to feed the network. This can be done in two alternative and complementary manners: (1) by a separate PCA of the calibration matrix **X**, estimating the number of principal components needed to capture most of the spectral variance, and (2) tuning the number of latent variables for a PCR analysis of the calibration data by cross validation. It might be argued here that principal component analysis is a linear technique, and that the scores generated by a linear technique might not be suitable for feeding a non-linear model. However, what is relevant here is that the PCA scores are adequate surrogate variables replacing the original spectra, capturing as much variance as possible. The non-linear relationship occurs between scores and concentrations, and this is the relationship modeled by the neural network.

In the framework of MVC1, the protocol described in Table 12.1 has been adopted. The tuning of the number of hidden neurons is called *forward selection*, because neurons are added one by one, until a minimum in a suitable error indicator is found.

Table 12.1 Optimization of an RBF network in MVC1

Step	Activity
1	A single hidden neuron is considered ($G = 1$)
2	A **C** matrix of Gaussian centers (size $G \times A$) is built with G randomly selected rows of the score matrix
3	A range of Gaussian widths (σ) is scanned in a proper range
4	A range of ridge regression parameters (λ) is scanned in a proper range
5	For each pair of values of σ and λ, the vector of weights **w** is estimated by ridge regression as in Eq. (12.11)
6	For each pair of values of σ and λ, an error indicator is computed, penalizing by the number of adjustable parameters the RMSEP value (computed by comparing nominal and predicted concentrations for the training samples)
7	Repeat the steps from 1 to 7 adding one hidden neuron at a time
8	The optimum values of G, σ, and λ correspond to the minimum error indicator

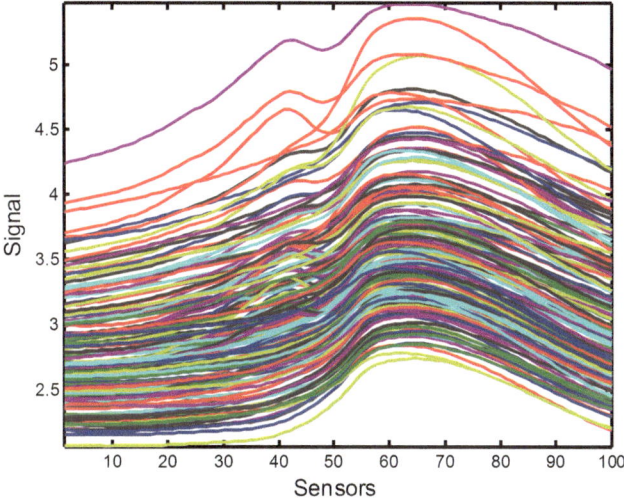

Fig. 12.3 NIR spectra of 170 meat samples, employed to build multivariate models for the non-invasive determination of the contents of fat, moisture, and protein. The data were recorded on a Tecator Infratec Food and Feed Analyzer working in the wavelength range 850–1050 nm, and are available at http://lib.stat.cmu.edu/datasets/tecator

12.9   A Real Case

This system involves the non-invasive determination of three quality parameters in meat samples, based on NIR spectral measurements in the range from 850 to 1050 nm (Borggaard and Thodberg 1992). A total of 170 samples were available for calibration (Fig. 12.3), and 70 for validation, with properties in the following ranges (%): fat, 32.8–76.6; moisture, 0.9–58.5; and protein, 8.8–23.2. We analyze in

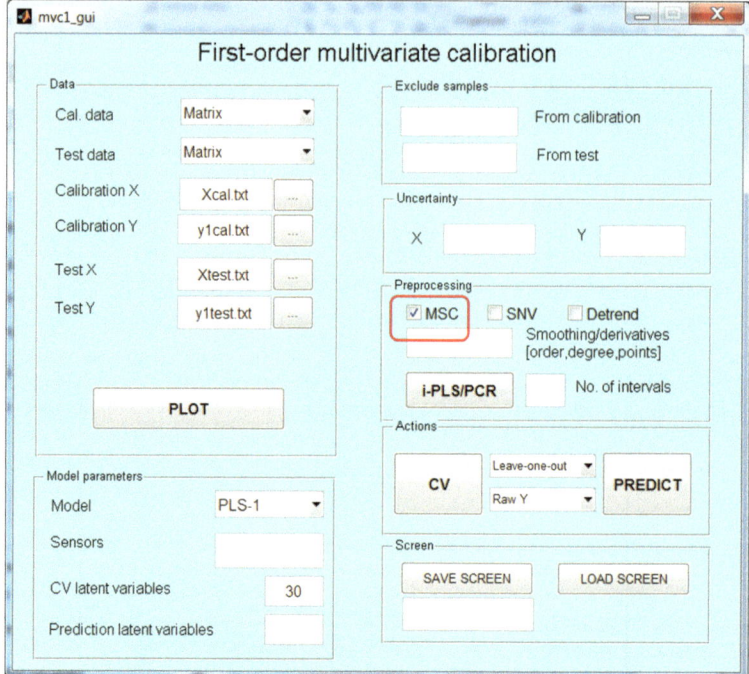

Fig. 12.4 MVC1 main screen, prepared to perform cross validation with the PLS-1 model, up to a maximum of 30 latent variables. The red box shows that MSC is employed as pre-processing filter. The system corresponds to NIR spectra for measuring quality parameters of meat samples, in this case fat

detail the determination of the fat content in these meat samples using the MVC1 software and an RBF approach, in comparison with classical PLS.

The experimental data set is available on the internet at http://lib.stat.cmu.edu/ datasets/tecator, and is contained in the folder named *Parameters in meat* accompanying the MVC1 software. We recommend the reader to examine the files contained in the latter folder, and to read the documentation which provides help and additional information on this system.

Figure 12.4 shows the main screen of MVC1, prepared to perform LOO-type cross validation using a PLS-1 model for the property fat, with a maximum of 30 latent variables. Notice that MSC is being used for pre-processing the NIR spectral data, due to the fact that solid or semi-solid materials introduce background dispersion signals which should be removed for better performance. Cross-validation analysis shows that the optimum number of latent variables is 3, and that the relationship between predicted fat content and nominal values is, at least visually, non-linear (Fig. 12.5).

Having established that the optimum number of latent variables is 3, Fig. 12.6 shows how the MVC1 screen is adapted to build the PLS-1 model for fat content. Prediction in the validation samples (Fig. 12.7) clearly indicates, by visual

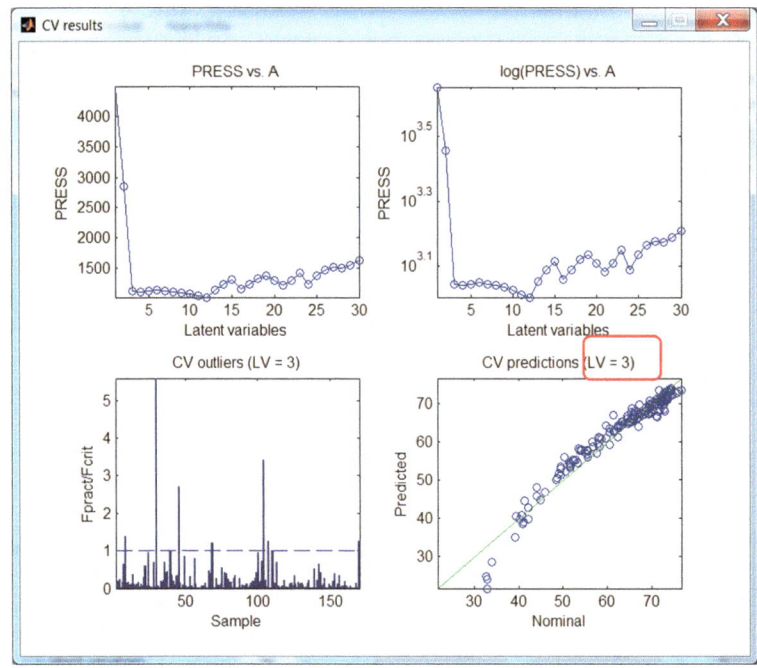

Fig. 12.5 MVC1 plots of cross-validation results for 3 latent variables (red box) in the PLS-1 determination of fat in meat samples

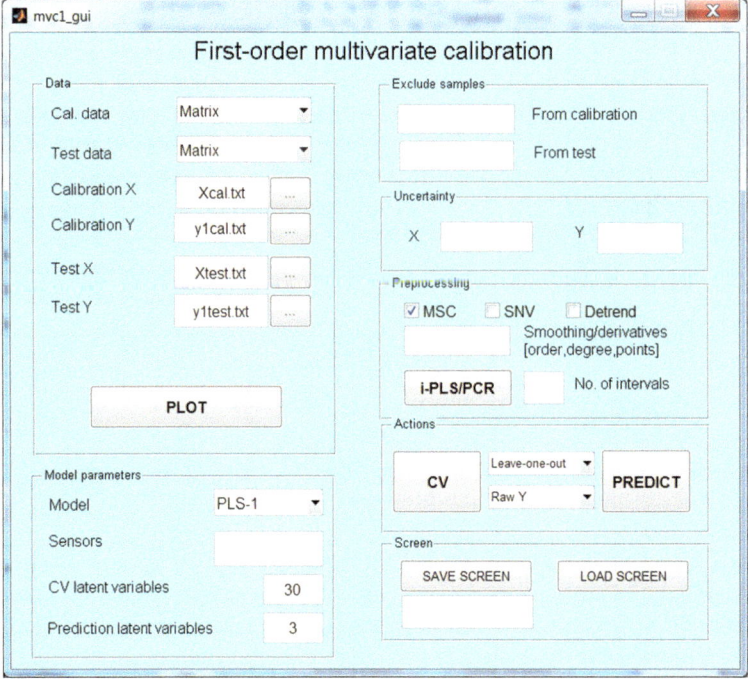

Fig. 12.6 MVC1 main screen, prepared to build a PLS-1 model with 3 latent variables for the determination of fat in meat samples

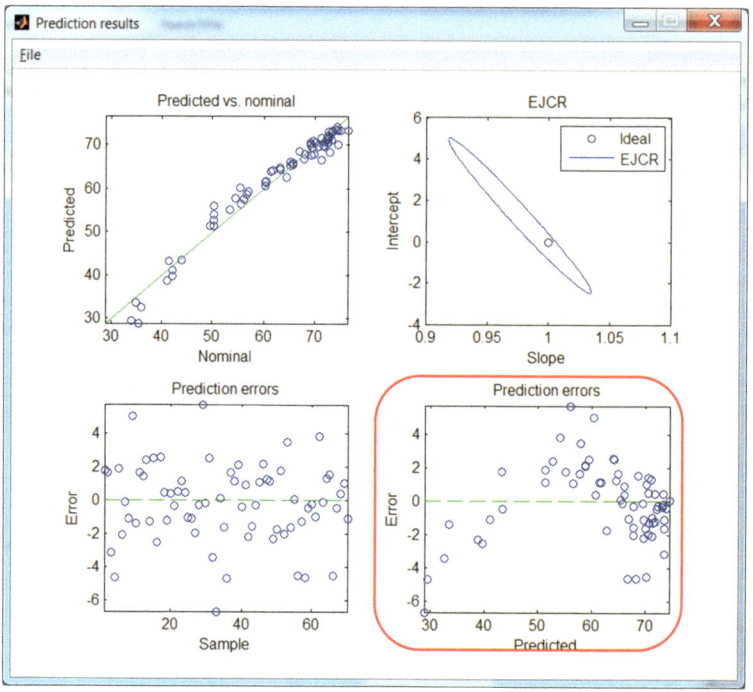

Fig. 12.7 MVC1 prediction plots for the PLS-1 model. The red box shows the prediction errors vs. predicted concentration

inspection, that the relation between predicted and nominal values is non-linear. Figure 12.8 shows the statistical results for the prediction phase: RMSEP = 2.3%, REP = 3.7%. Notice that the Durbin–Watson indicator for the correlation among prediction residuals is 0.90, with a very low associated probability, indicating that non-linearity is indeed significant.

Given the non-linear nature of the signal–concentration relationship, the prediction results could be improved by moving to a non-linear approach such as those based on RBF neural networks. To apply the latter model requires one to set a suitable number of latent variables capturing the spectral calibration variance. A separate PCA study is required for estimating the number of latent variables. A hint that this number is larger than 3 is provided by an LOO-type cross-validation analysis of the PCR model, which yields 13 latent variables as optimum. We may even use a larger number; the aim is to capture as much variance as possible, and we do not want to leave PCs out of the model.

Figure 12.9 shows the pertinent MVC1 screen, where 20 latent variables were selected as input of the network. RBF prediction leads to significantly better analytical results (Fig. 12.10), where it is apparent that the correlation among successive residuals is considerably smaller. In particular, the prediction statistics (Fig. 12.11) looks better than the one for PLS-1, with RMSEP = 0.84% and REP = 1.3%.

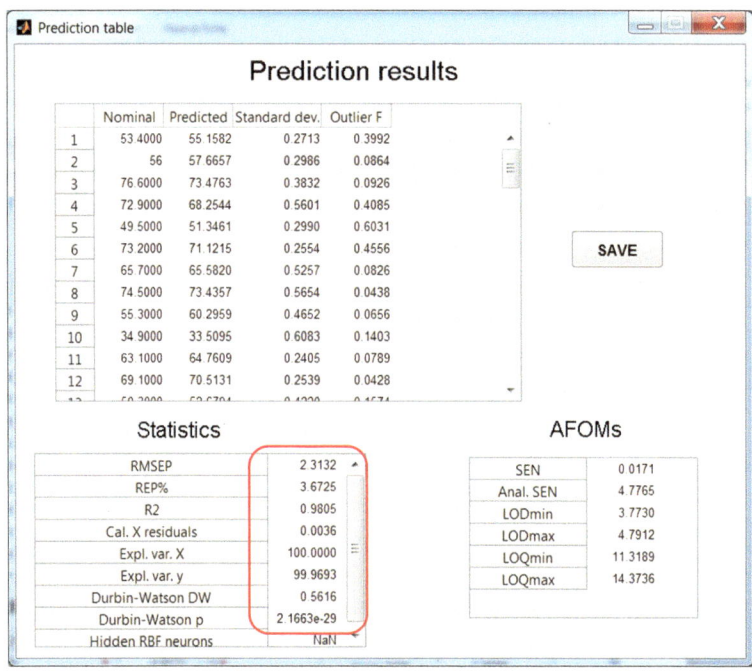

Fig. 12.8 MVC1 table with prediction results. The red box shows the statistical indicators for the prediction phase

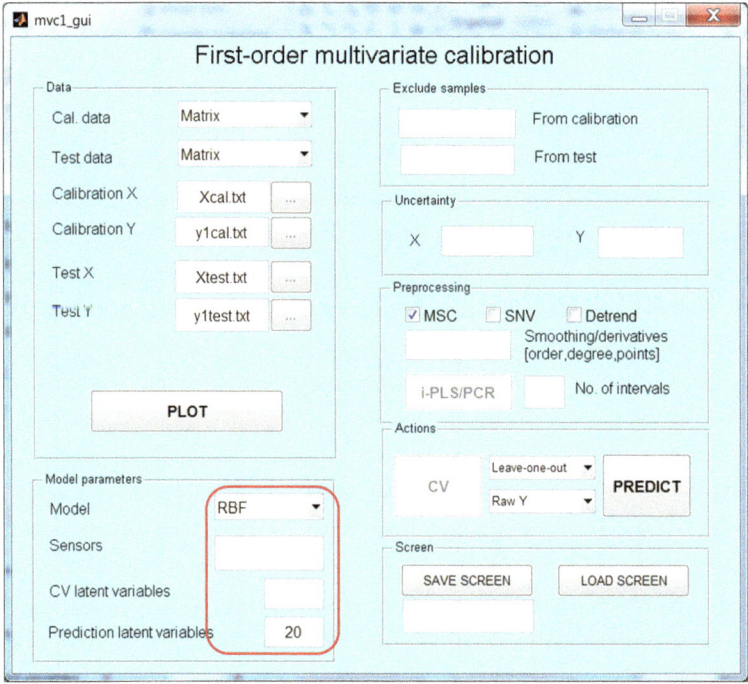

Fig. 12.9 MVC1 main screen, prepared to train an RBF neural network with 20 scores as input, corresponding to an input layer with 20 neurons (red box). The analytical system is the same as in Fig. 12.8. MSC is employed as pre-processing filter

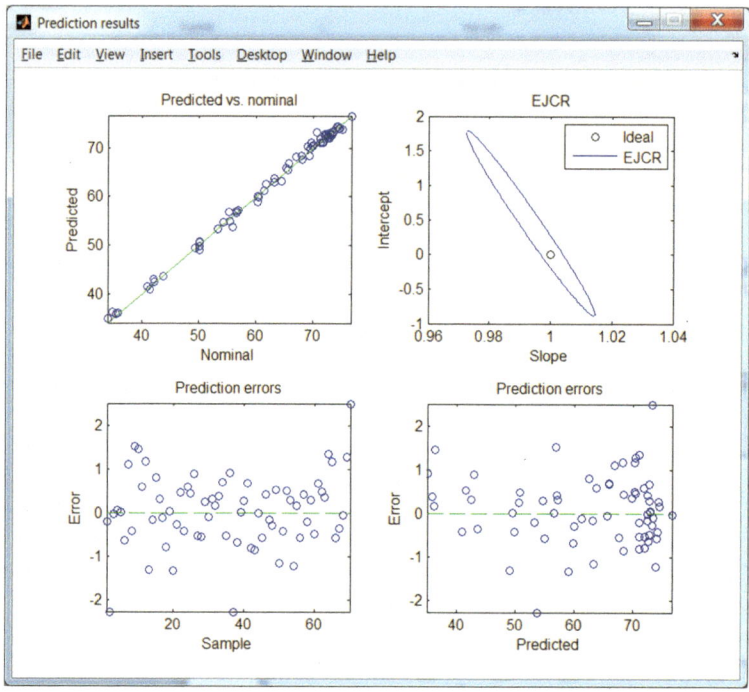

Fig. 12.10 MVC1 plots of fat prediction results in meat samples with the trained RBF

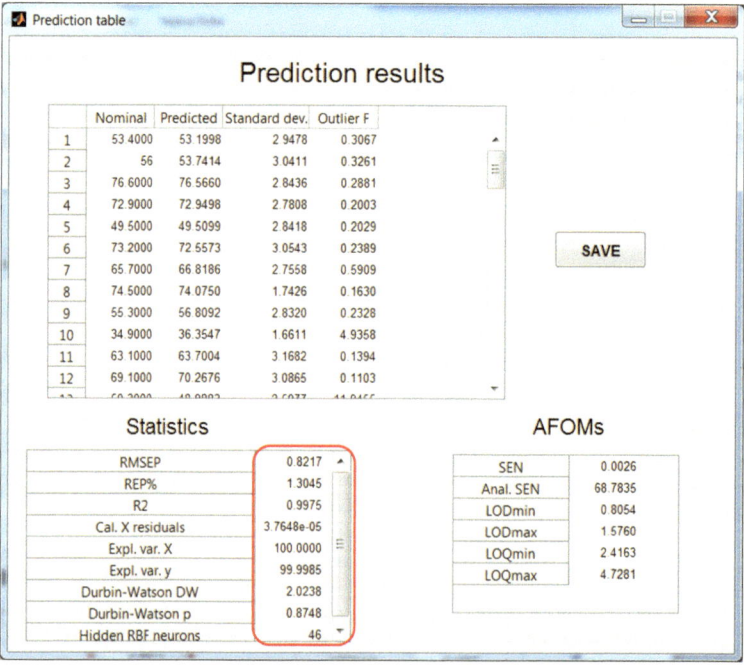

Fig. 12.11 MVC1 table of prediction results for the RBF model. The red box shows the statistical indicators for the RBF prediction phase

Table 12.2 Results for the determination of quality parameter of meat samples by PLS-1 and RBF calibrations

	Fat/%	Moisture/%	Protein/%
No. of calibration samples	170		
No. of test samples	70		
Property range/%	32.8–76.6	0.9–58.5	8.8–23.2
Mean calibration value/%	63.0	18.3	17.7
	PLS-1 calibration		
Number of latent variables	3	11	10
RMSEP/%	2.3	2.8	0.8
REP/%	3.7	15	4.3
Durbin–Watson DW	0.90	1.17	1.68
Durbin–Watson p	$\ll 0.05$	$\ll 0.05$	0.035
	RBF calibration		
ANN architecture[a]	20-46-1	20-61-1	20-53-1
RMSEP/%	0.82	0.65	0.54
REP/%	1.3	3.5	3.1
Durbin–Watson DW	2.02	1.93	2.10
Durbin–Watson p	0.87	0.66	0.49

[a]The architecture is reported as number of input-hidden-output neurons

The Durbin–Watson indicator is 2.1 with a probability $p = 0.40$, clearly indicating non-significant correlations. This confirms that the prediction improvement using the neural network approach is due to an adequate mapping of the non-linear signal–concentration relation.

This data set has additional quality parameters to calibrate: moisture and protein content. The reader will be able to verify the results provided in Table 12.2 (Exercise 2 in Sect. 12.11). It is apparent that RBF calibrations provide reasonably good results, with non-significant correlations in prediction residuals, unlike PLS-1. The only exception appears to be the determination of the protein content, for which the Durbin–Watson probability is close to 0.05. Indeed, PLS-1 calibration with 10 latent variables for protein yields RMSEP = 0.76%, REP = 4.3%, close to the RBF result. This may indicate that the relationship between NIR signals and protein content in these meat samples is only slightly non-linear, and that the improvement in going from PLS-1 to RBF may only be marginal.

12.10 Figures of Merit

We recall that in PLS calibration, the vector of regression coefficients $\mathbf{b}_{PLS}$ can be used to estimate the sensitivity:

$$\mathrm{SEN}_{\mathrm{PLS}} = \frac{1}{\sqrt{\sum\limits_{j=1}^{J} b_{j\mathrm{PLS}}^2}} \tag{12.16}$$

Interestingly, there is a simple equivalent to Eq. (12.16) in RBF calibration (Allegrini and Olivieri 2016). Starting from the general prediction Eq. (12.15) for RBFs, and applying uncertainty propagation, it is possible to derive an expression formally analogous to Eq. (12.16), as a function of an analyte and sample dependent $\mathbf{b}_{\mathrm{RBF}}$ vector whose generic elements $b_{\mathrm{RBF}}(j)$ are given by:

$$b_{\mathrm{RBF}}(j) = -\sum_{g=1}^{G} w_g \left[\sum_{a=1}^{A} \frac{2(t_a - c_{ga}) u_{ja}}{\sigma^2} \right] \exp\left(-\frac{\sum\limits_{a=1}^{A} (t_a - c_{ga})^2}{\sigma^2} \right) \tag{12.17}$$

where u_{ja} are the elements of the loading vectors rendered by PCA of the raw data matrix, and the remaining symbols have the same meaning as above. The sensitivity for the network calibration is thus:

$$\mathrm{SEN}_{\mathrm{RBF}} = \frac{1}{\sqrt{\sum\limits_{j=1}^{J} b_{j\mathrm{RBF}}^2}} \tag{12.18}$$

It is worth mentioning that $\mathbf{b}_{\mathrm{RBF}}$ coefficients, in contrast to PLS regression coefficients, are not employed for analyte prediction. They are sensitivity coefficients which are sample specific, meaning that each sample will be characterized by a sensitivity value. This can be understood considering that the slope of the signal–concentration relationship in a non-linear system is not constant. The sensitivity can be reported as an average value, ranging from a minimum to a maximum.

Equations (12.17) and (12.18) make it possible to develop expressions for the sample dependent prediction uncertainty and detection limit. However, the correct estimation of the latter parameters is still a subject of debate and intense research (Allegrini and Olivieri 2016).

Figure 12.12 shows additional model plots. The one highlighted within the red box shows the vectors of $\mathbf{b}_{\mathrm{RBF}}$ coefficients for all the calibration samples, which are employed to estimate the sensitivities according to Eq. (12.18). The sensitivity is specific for each sample, and thus MVC1 provides the average value <SEN$_{\mathrm{RBF}}$>. The obtained values are shown in Table 12.3 along with the minimum and maximum sensitivity parameter for each of the three RBF calibrations.

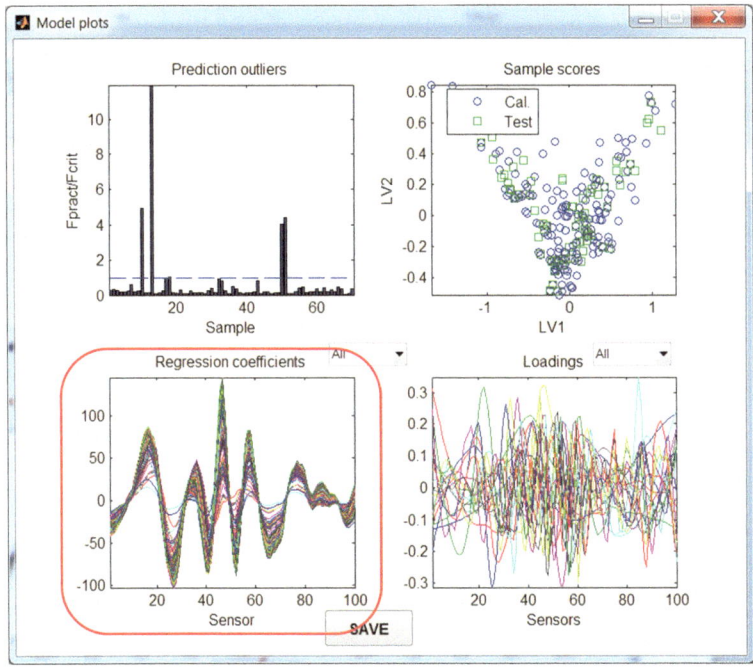

Fig. 12.12 MVC1 model plots. The red box shows the vectors of RBF coefficients which are useful to estimate the sensitivity of the model

Table 12.3 Figures of merit for the determination of quality parameter of meat samples by RBF calibration

	Fat/%	Moisture/%	Protein/%
Average $\langle SEN_{RBF}\rangle$	0.0038	0.0051	0.0022
Minimum SEN_{RBF}	0.0017	0.0037	0.0015
Maximum SEN_{RBF}	0.049	0.030	0.022

12.11 Exercises

1. Explain whether the following statements are true or false:
   (a) The ANN calibration results are better than those from PLS, even when there is a linear relationship between signal and concentration.
   (b) ANN calibrations are to be preferred over PLS calibration in very complex samples, even for linear signal–concentration relationships.
   (c) ANN require fewer samples for building the calibration model.
   (d) ANN programs are very complex and time consuming.
   (e) The mathematics associated to ANN is difficult to interpret.
   (f) No figures of merit are available for ANN calibrations.

2. Process the data set called *Parameters in meat* with PLS-1 and RBF using the MVC1 software, and reproduce the results shown in Table 12.2. Apply MSC as pre-processing. Check whether the system is non-linear for the three calibrated parameters, on the basis of the Durbin–Watson indicators for PLS-1 calibration.
3. Process the data set called *Parameters in corn* with PLS-1 and RBF using the MVC1 software and collect the results in a table similar to Table 12.2. Apply MSC as pre-processing. Check whether the system is non-linear for the four calibrated parameters, on the basis of the Durbin–Watson indicators for PLS-1 calibration. Are there significant differences between the linear and non-linear calibrations?

References

Allegrini, F., Olivieri, A.C.: Sensitivity, prediction uncertainty, and detection limit for artificial neural network calibrations. Anal. Chem. **88**, 7807–7812 (2016)

Baffin, G., Martin, E.B., Morris, A.J.: Non-linear projection to latent structures revisited: the quadratic PLS algorithm. Comput. Chem. Eng. **23**, 395–411 (1999)

Borggaard, C., Thodberg, H.H.: Optimal minimal neural interpretation of spectra. Anal. Chem. **64**, 545–551 (1992)

Centner, V., de Noord, O.E., Massart, D.L.: Detection of nonlinearity in multivariate calibration. Anal. Chim. Acta. **376**, 153–168 (1998)

Despagne, F., Massart, D.L.: Neural networks in multivariate calibration. Analyst. **123**, 157R–178R (1998)

Durbin, J., Watson, G.S.: Testing for serial correlation in least squares regression I. Biometrika. **37**, 409–428 (1950)

Haykin, S.: Neural Networks. A Comprehensive Foundation, 2nd edn. Prentice-Hall, Upper Saddle River, NJ (1999)

Ni, W., Nørgaard, L., Mørupc, M.: Non-linear calibration models for near infrared spectroscopy. Anal. Chim. Acta. **813**, 1–14 (2014)

Orr, M.J.L.: Introduction to radial basis function networks. Centre for Cognitive Science, Edinburgh University, Edinburgh, Scotland, pp. 1–67 (1996)

Zupan, J., Gasteiger, J.: Neural Networks in Chemistry and Drug Design, 2nd edn. Wiley VCH, Weinheim (1999)

Solutions to Exercises

13

Abstract

Solutions to the exercises provided at the end of each chapter are worked out in detail.

13.1 Chapter 2

1. (a) Alternative 1: $\mathbf{Y}^T\mathbf{Y} = \begin{bmatrix} 1 & 0 \\ 0 & 1 \end{bmatrix}^T \begin{bmatrix} 1 & 0 \\ 0 & 1 \end{bmatrix} = \begin{bmatrix} 1 & 0 \\ 0 & 1 \end{bmatrix} \begin{bmatrix} 1 & 0 \\ 0 & 1 \end{bmatrix} = \begin{bmatrix} 1 & 0 \\ 0 & 1 \end{bmatrix}$

   Alternative 2: $\mathbf{Y}^T\mathbf{Y} = \begin{bmatrix} 1 & 1 \\ 2 & 2 \end{bmatrix}^T \begin{bmatrix} 1 & 1 \\ 2 & 2 \end{bmatrix} = \begin{bmatrix} 1 & 2 \\ 1 & 2 \end{bmatrix} \begin{bmatrix} 1 & 1 \\ 2 & 2 \end{bmatrix} = \begin{bmatrix} 5 & 5 \\ 5 & 5 \end{bmatrix}$

   Alternative 3: $\mathbf{Y}^T\mathbf{Y} = \begin{bmatrix} 1 & 2 \\ 2 & 1 \end{bmatrix}^T \begin{bmatrix} 1 & 2 \\ 2 & 1 \end{bmatrix} = \begin{bmatrix} 1 & 2 \\ 2 & 1 \end{bmatrix} \begin{bmatrix} 1 & 2 \\ 2 & 1 \end{bmatrix} = \begin{bmatrix} 5 & 4 \\ 4 & 5 \end{bmatrix}$.

   (b) Alternative 1: $\det(\mathbf{Y}^T\mathbf{Y}) = 1$

   Alternative 2: $\det(\mathbf{Y}^T\mathbf{Y}) = 0$
   Alternative 3: $\det(\mathbf{Y}^T\mathbf{Y}) = 9$.

   (c) Alternative 2 should be discarded because the determinant is zero and the matrix cannot be inverted. Qualitatively, the two samples have the same analyte concentrations; they are, in fact, duplicates of the same sample and therefore they are not independent. Calibrating for two analytes requires at least two independent samples.

   On the other hand, alternatives 1 and 3 do not present this problem. Which one to choose? Alternative 1 does not involve mixtures, but pure analyte solutions, whereas alternative 3 involves two mixtures with different analyte proportions. In general, it is preferable to choose alternative 3, which may be more representative of future samples, where the analytes may show slightly different spectra in comparison with pure analyte spectra.

© Springer Nature Switzerland AG 2018
A. C. Olivieri, *Introduction to Multivariate Calibration*,
https://doi.org/10.1007/978-3-319-97097-4_13

2. (a) The determination of the quality of chicken meat cannot be performed using a CLS model, because this would require to know all chemical constituents of the meat, and the mathematical relationship between the quality parameters and the chemical components.

(b) A constant background can be removed by measuring a blank sample, and then subtracting the background signal from all samples, so that if the two pure chemical constituents are known, the CLS model could be applied.

(c) A variable background cannot be easily removed, so that, in general, the CLS model will not be useful in this particular case.

3. (a) The spectral overlapping increases in the order B < A < C, and the determinant of the matrix $(\mathbf{S}^T\mathbf{S})$ decreases in the same order.

(b) The most favorable situation for inverting the above matrix and estimating analyte concentrations is B.

13.2 Chapter 3

1. (a) For a CLS model, in the prediction phase the vector of analyte concentrations is given by:
$$\mathbf{y} = (\mathbf{S}^T\mathbf{S})^{-1}\mathbf{S}^T\mathbf{x} = \mathbf{S}^+\mathbf{x}$$
If the number of analytes is equal to the total number of constituents, the above equation is analogous to the ILS expression for all analytes:
$$\mathbf{y} = \mathbf{B}\mathbf{x}$$
This shows that the matrix of regression coefficients of the ILS model is $\mathbf{S}^+$, if all constituent pure spectra are known.

(b) Inversion of the $N \times N$ $(\mathbf{S}^T\,\mathbf{S})$ matrix requires $J > N$ and low correlation among the columns of $\mathbf{S}$.

(c) There are four constituents, low spectral overlapping and no background signal, so that the number of wavelengths selected by SPA is equal to the number of constituents.

2. Not necessarily, but this is an indication that various chemical constituents are involved in this determination.

3. (a) moisture $= k(x_{1940} - x_{2080}) = kx_{1940} - kx_{2080} = [k \; -k]\begin{bmatrix} x_{1940} \\ x_{2080} \end{bmatrix} = \mathbf{b}^T_{\text{moisture}}\mathbf{x}$

(b) $\mathbf{b}_{\text{moisture}} = \begin{bmatrix} k \\ -k \end{bmatrix}$

4. (a) Registering the NIR spectra of a large number of chicken meat samples, measuring their quality parameters, selecting the working wavelengths with e.g., SPA, and building an ILS model at those wavelengths.

(b) Registering the NIR spectra of a large number of two-constituent mixtures of known concentrations in the presence of the variable background signal, selecting the working wavelengths with e.g., SPA, and building an ILS model at those wavelengths.

5. (a) $\mathbf{v}_n = (\mathbf{T}^T\mathbf{T})^{-1}\mathbf{T}^T\mathbf{y}_n$

   (b) If the size of $\mathbf{T}$ is $I \times A$, where I is the number of calibration samples and A is the size of the compressed and truncated matrix $\mathbf{T}$, then the requirement for inverting the square matrix $(\mathbf{T}^T\mathbf{T})$ of size $A \times A$ is that $I > A$.

6. Metal ions should be bonded to organic matter, so that the content of organic matter, determined by NIR and chemometrics, is proportional to the metal content.

7. In ILS, the function to be minimized is:
$$f_{ILS} = (\mathbf{X}^T\mathbf{b}_n - \mathbf{y}_n)^T(\mathbf{X}^T\mathbf{b}_n - \mathbf{y}_n)$$
The minimum in f_{ILS} is found by setting to zero the first derivative with respect to $\mathbf{b}_n$. The derivative is a vector of the same size as $\mathbf{y}_n$, all whose elements should be zero, represented by the symbol "$\mathbf{0}$":
$$df_{ILS}/d\mathbf{b}_n = 2\mathbf{X}(\mathbf{X}^T\mathbf{b}_n - \mathbf{y}_n) = \mathbf{0}$$
which leads to:
$$\mathbf{X}^T\mathbf{b}_n - \mathbf{y}_n = \mathbf{0}$$
$$\mathbf{X}^T\mathbf{b}_n = \mathbf{y}_n$$
$$\mathbf{X}\,\mathbf{X}^T\mathbf{b}_n = \mathbf{X}\mathbf{y}_n$$
$$\mathbf{b}_n = (\mathbf{X}\,\mathbf{X}^T)^{-1}\mathbf{X}\mathbf{y}_n$$
In RR, the function to be minimized is:
$$f_{RR} = (\mathbf{X}^T\mathbf{b}_n - \mathbf{y}_n)^T(\mathbf{X}^T\mathbf{b}_n - \mathbf{y}_n) + \lambda\mathbf{b}_n^T\mathbf{b}_n$$
The minimum in f_{RR} is found by setting to zero the first derivative:
$$df_{RR}/d\mathbf{b}_n = 2\mathbf{X}(\mathbf{X}^T\mathbf{b}_n - \mathbf{y}_n) + 2\lambda\mathbf{b}_n = \mathbf{0}$$
which leads to:
$$\mathbf{X}(\mathbf{X}^T\mathbf{b}_n - \mathbf{y}_n) + \lambda\mathbf{b}_n = \mathbf{0}$$
$$\mathbf{X}\mathbf{X}^T\mathbf{b}_n + \lambda\mathbf{b}_n = \mathbf{X}\,\mathbf{y}_n$$
$$(\mathbf{X}\mathbf{X}^T + \lambda\mathbf{I})\mathbf{b}_n = \mathbf{X}\mathbf{y}_n$$
$$\mathbf{b}_n = (\mathbf{X}\,\mathbf{X}^T + \lambda\mathbf{I})^{-1}\mathbf{X}\mathbf{y}_n$$

13.3 Chapter 4

1. (a) In the uncentered data, only 1.0% of the variance is explained by the second score, which is the one mainly responsible for classification. On the other hand, for centered data, the explained variance by the second score increased to 18.6% by removing the effect of the mean spectrum. More importance is now given to the effect of the discriminating spectral shoulder at sensor 30.

   (b) In the centered data, the second loading has a smaller contribution from the sensor 60, which represents the mean spectrum, and a larger contribution from the sensor 30, which represents the contribution of the discriminating spectral shoulder.

   (c) In the score–score plot for centered data, the samples are clearly discriminated according to the value of the second score: positive values, green samples, negative values, blue samples.

13.4   Chapter 5

1. It is likely that PCR, using full spectral information, is able to model the sample properties with higher efficiency than ILS.
2. If we accept the views of this book, answer (a) is the appropriate one to the question.

13.5   Chapter 6

1. (a) False. You cannot have so many latent variables when only two analytes are responsive in a solution system.
   (b) May be true if the additional latent variables model dispersion effects.
   (c) True. After removing the background signals, the number of latent variables tends to decrease.
   (d) False. Three responsive analytes in solution require at least three latent variables.

13.6   Chapter 7

1. (a) From Eq. (7.1) of Chapter 7:
   $$\mathbf{y}_n = \mathbf{T}_A \mathbf{v}_n$$
   $$\mathbf{v}_n = (\mathbf{T}_A^T \mathbf{T}_A)^{-1} \mathbf{T}_A^T \mathbf{y}_n$$
   $$\mathbf{v}_n = \mathbf{T}_A^+ \mathbf{y}_n$$
   (b) $\mathbf{T}_A^+ = (\mathbf{T}_A^T \mathbf{T}_A)^{-1} \mathbf{T}_A^T$

2. The number of independent equations is equal to the number of calibration samples I. The number of unknowns is equal to the number of latent variables A. The requirements for solving the equation is that $A < I$, and that the columns of $\mathbf{T}_A$ are not correlated.
3. The three vectors are as follows:

$$
\mathbf{y}_1 = \begin{bmatrix} 1 \\ 1 \\ 1 \\ 1 \\ 1 \\ 0 \\ 0 \\ 0 \\ 0 \\ 0 \\ 0 \\ 0 \\ 0 \\ 0 \\ 0 \end{bmatrix}
\quad
\mathbf{y}_2 = \begin{bmatrix} 0 \\ 0 \\ 0 \\ 0 \\ 0 \\ 1 \\ 1 \\ 1 \\ 1 \\ 0 \\ 0 \\ 0 \\ 0 \\ 0 \\ 0 \end{bmatrix}
\quad
\mathbf{y}_3 = \begin{bmatrix} 0 \\ 0 \\ 0 \\ 0 \\ 0 \\ 0 \\ 0 \\ 0 \\ 0 \\ 1 \\ 1 \\ 1 \\ 1 \\ 1 \\ 1 \end{bmatrix}
$$

4. (a) Three latent variables, due to the three chemical constituents.
   (b) More than three latent variables, due to the three chemical constituents and the dispersion of the NIR radiation.
5. (a) The interferent will be flagged as an outlier because the signal from the interferent cannot be modeled by a calibration set where it is absent.
   (b) It is likely that the bias will be very small, because the elements of the regression vector at the wavelengths where the interferent absorbs are very small.
   (c) The interferent will be flagged as an outlier, and there will be considerable bias in the analyte prediction, because the elements of the regression vector at the wavelengths where the interferent absorbs are significant.
6. Samples with large spectral residuals are more problematic than samples having unusual constituent concentrations, because the latter concentrations may be outside the calibration range, but the former are due to uncalibrated new constituents. If the signal–concentration relationship is linear, high constituent concentrations should not be a problem. New constituents, however, demand expanding the calibration set with additional representative samples.
7. (a) No. Values in the first row should be reported as 0.19, 0.15, and 0.5, respectively.
   (b) Visual inspection indicates that ILS is definitely the worst model. PLS-1 and PLS-2 appear to be equivalent, most probably because there are no correlations among the analyte calibration concentrations. PLS-1 should be the model of choice.

13.7   Chapter 8

1. (a) The sub-region from 200 to 500 sensors appears to be the best for calibration.
   (b) Calibrating with the full spectra requires more latent variables, to model additional phenomena due to saturation of the detector and presence of other chemical constituents in comparison with the sensor-selected region. The RMSEP is smaller in the selected sub region because it may be more informative regarding the octane number.

13.8   Chapter 9

1. The smallest REP corresponds to MSC, but there are almost no differences with the one obtained by applying no pre-processing. The samples are liquid, and no dispersion of the radiation is expected.

13.9   Chapter 10

1. (a) In general, yes, unless the uncertainty in prediction is controlled by the errors
   in nominal or reference values for the calibration samples.
   (b) Same answer as (a).
   (c) Yes, because $LOQ = 3$ LOD.
2. No. Predictions have too many significant figures. The correct report is shown in
   Table 13.1:

Table 13.1 A report with the correct number of significant figures

Sample	Nominal	Prediction
1	0.10	0.10(1)
2	0.20	0.21(2)
3	0.30	0.30(1)
4	0.40	0.42(1)
5	0.50	0.51(2)
RMSEP		0.011
REP		36%
SEN		1.16
LOD		0.012
LOQ		0.04

13.10   Chapter 11

1. The example data set *Tetracycline in serum* is analyzed following the suggested
   protocol:
   (a) No saturated spectral regions appear in the spectra
   (b) LOO cross validation with a maximum of 25 latent variables is performed
   (there are 50 calibration samples). Four latent variables is optimum for the
   PLS-1 model, with RMSECV $= 0.14$ units (satisfactory since the average
   calibration concentration is ca. 2 units, 7% of relative error for human serum
   samples).
   (c) No outliers appear in cross validation.
   (d) and (e) Prediction results with PLS-1 and 4 latent variables are reasonably
   good: RMSEP $= 0.07$ and REP $= 3.8\%$. Figures of merit: the LOD is also
   reasonably low, from 0.14 to 0.22 units.
   (f) No significant outliers in prediction.
   (g) No truly unknown samples available.
2. The example data set *Octane in gasolines* is analyzed following the suggested
   protocol:
   (a) Saturated spectral regions appear in the spectra below sensor No. 200
   (b) LOO cross validation with a maximum of 24 latent variables is performed
   (there are 48 calibration samples) in the spectral region from 200 to

500 sensors. Five latent variables is optimum for the PLS-1 model, with RMSECV = 0.6 units, but there are two clear outliers: samples No. 24 and 25. Excluding these two samples and repeating the LOO cross validation leads to four latent variables and RMSECV = 0.4 units.

(c) No significant outliers appear in cross validation in the spectral region 200–500 and excluding calibration samples 24 and 25.

(d) and (e) Prediction results with PLS-1 and 4 latent variables are reasonably good: RMSEP = 0.28 and REP = 0.29%. Figures of merit: the LOD does not have a meaning in this case. However, the analytical sensitivity (22 units) provides an idea of the minimum difference in property that can be appreciated, as 1/22 = ca. 0.05 units.

(f) No significant outliers in prediction.

(g) No truly unknown samples available.

13.11 Chapter 12

1. (a) Not necessarily.
   (b) Not necessarily.
   (c) Not necessarily.
   (d) They are not very complex, and some (RBF) are not time consuming.
   (e) See the present chapter. Is it?
   (f) They are for RBF, at least the sensitivity can be readily estimated.
2. Reproduce Table 13.1.
3. Processing the data with MVC1 using both PLS-1 and RBF (MSC pre-processing was used) leads to Table 13.2.

Table 13.2 PLS-1 and RBF results for the determination of corn seed quality parameters

	Moisture (%)	Oil (%)	Protein (%)	Starch (%)
PLS-1 calibration[a]				
Number of latent variables	21	19	19	17
RMSEP (%)	0.21	0.07	0.13	0.18
REP (%)	2.0	2.0	1.5	0.27
Durbin–Watson *DW*	1.88	2.41	2.88	1.24
Durbin–Watson *p*	0.57	0.17	0.001	0.003
RBF calibration				
ANN architecture[b]	20-13-1	20-9-1	20-21-1	20-25-1
RMSEP (%)	0.19	0.14	0.16	0.59
REP (%)	1.9	4.1	1.8	0.91
Durbin–Watson *DW*	2.18	1.72	2.48	2.07
Durbin–Watson *p*	0.61	0.25	0.11	0.89

[a]A maximum of 25 latent variables was employed for LOO cross validation
[b]The architecture is reported as number of input-hidden-output neurons

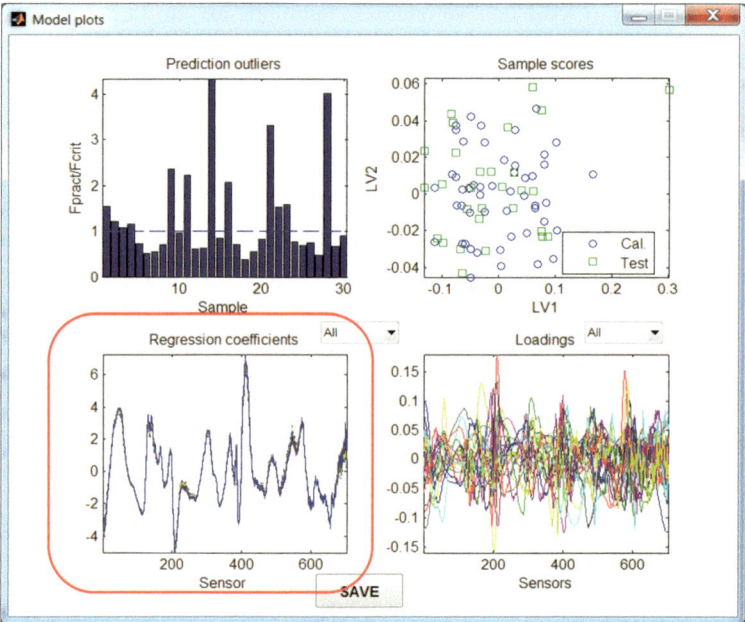

Fig. 13.1 MVC1 screen with the model plots for the RBF determination of corn seed quality parameters, in this case moisture. The plot of sensitivity RBF coefficients is inside the red box

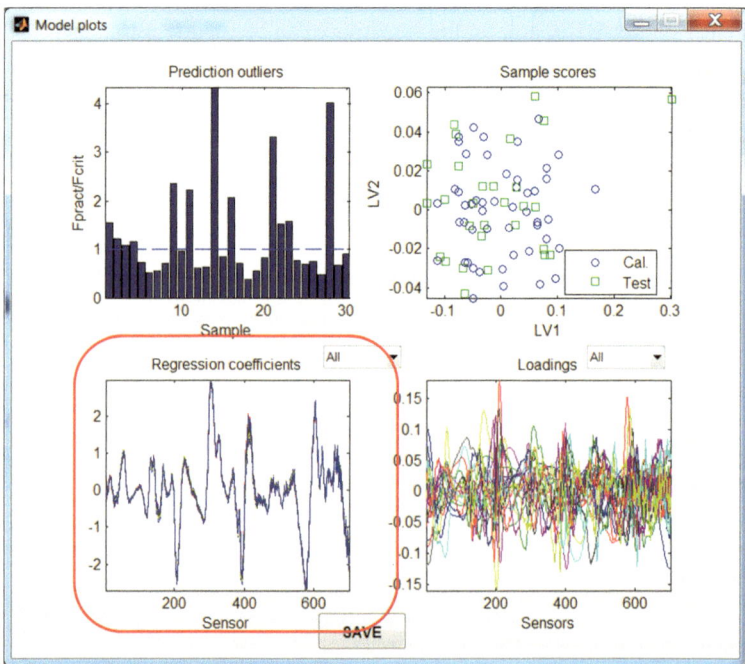

Fig. 13.2 MVC1 screen with the model plots for the RBF determination of corn seed quality parameters, in this case oil. The plot of sensitivity RBF coefficients is inside the red box

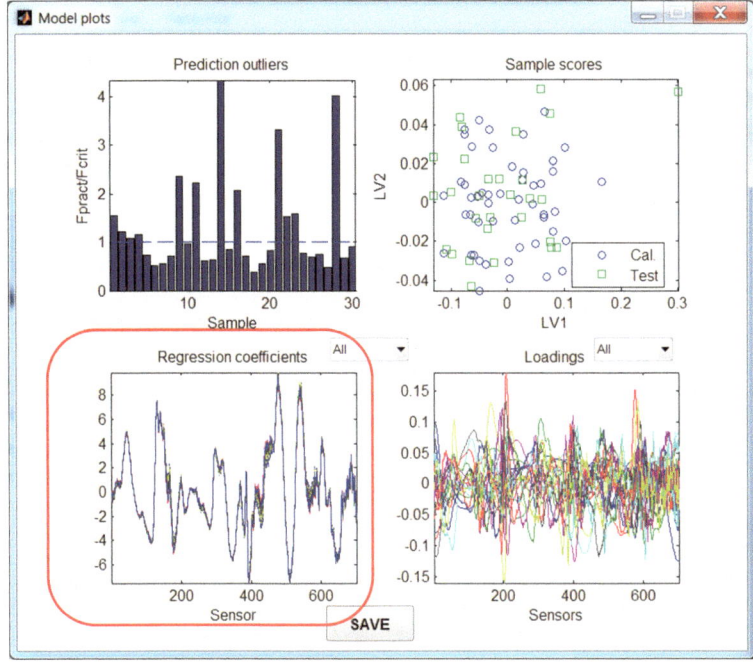

Fig. 13.3 MVC1 screen with the model plots for the RBF determination of corn seed quality parameters, in this case protein. The plot of sensitivity RBF coefficients is inside the red box

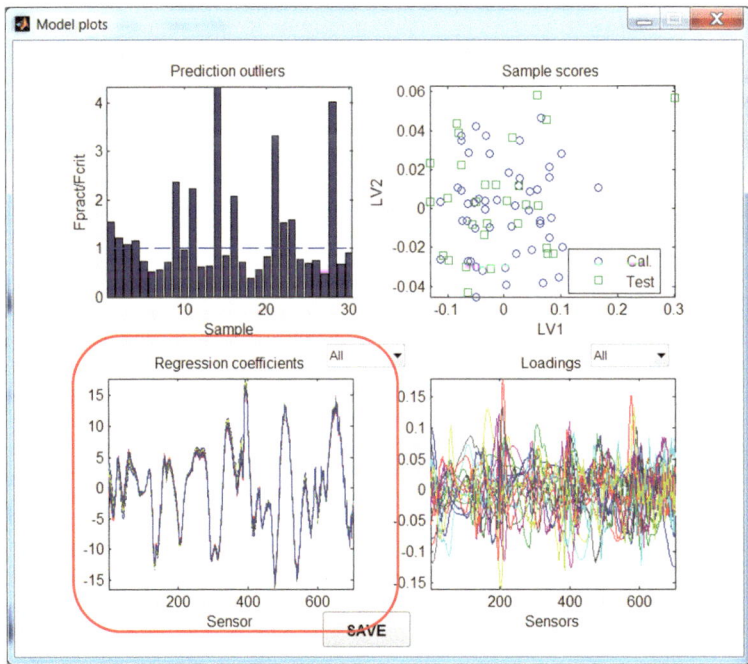

Fig. 13.4 MVC1 screen with the model plots for the RBF determination of corn seed quality parameters, in this case starch. The plot of sensitivity RBF coefficients is inside the red box

Conclusion: The system is linear. Even for protein and starch, where the probabilities associated to the DW parameters seem to indicate non-linearity, the predictions are of the same (or even better quality) than the use of the RBF model. All this indicates that for non-linearity, one should detect a very low Durbin–Watson probability, ($p < <0.05$) for significant non-linearity to occur. Values of 0.001–0.003 as in Table 13.2 should be taken with caution.

Another indication that the systems are linear is the RBF plot of the sensitivity vectors. If these vectors are similar, it is a strong indication that the sensitivity at all concentrations is the same, suggesting linearity. Figure 13.1 shows these plots for the four parameters of Table 13.2 (compare with Fig. 12.11 of Chap. 12 corresponding to a truly non-linear system) (For the remaining parameters, see Figs. 13.2, 13.3, and 13.4).

Back Matter

Let me see this book as if it were a tour. The visitor will stop at the sites that are most recommended by commercial agencies. The first one is CLS, a place that has been abandoned by almost all of its residents. However, it retains a high historical value, and prepares the visitor for the main objectives of the circuit. The second milestone of the tour is ILS, not too visited but still active. After some refurbishment here and there, ILS will be ready to receive new guests. It maintains its appeal, despite the passing of time and the discovery of new places of greater interest. ILS is part of the history of multivariate calibration, but may still be useful. Some users rely on ILS because of its simplicity.

Before arriving to the jewel of the tour, a stop at PCR is timely. Some people think that PCR is as important as PLS, and that in the future it could overshadow its splendor. But most researchers prefer PLS, for historical, emotional, or rational reasons. We do not know yet whether there are sufficient scientific reasons to assert that PLS is better than PCR. It may depend on what one understands by *better*. In any case, PLS is the most visited place today, the one with more stars according to travelers' opinions, and the most recommended one by touristic agencies. PLS has certainly more press than PCR.

ANN is then briefly visited before returning home, perhaps so quickly that its whole potential may not be appreciated. In the future, it may reach the same status as PLS, but today it is destined to receive no more than a glimpse. Brief, yes, but powerful enough so as to catch the travelers' interest.

Those who made the complete tour may once come back. For different causes: the wish of knowing more about a place of interest, the need of learning on the history, the present power or the future projections of some of the visited sites, or just by curiosity. They will all be welcome.

© Springer Nature Switzerland AG 2018
A. C. Olivieri, *Introduction to Multivariate Calibration*,
https://doi.org/10.1007/978-3-319-97097-4

Index

© Springer Nature Switzerland AG 2018
A. C. Olivieri, *Introduction to Multivariate Calibration*,
https://doi.org/10.1007/978-3-319-97097-4